Evolution from Cellular to Social Scales

NATO Science for Peace and Security Series

This Series presents the results of scientific meetings supported under the NATO Programme: Science for Peace and Security (SPS).

The NATO SPS Programme supports meetings in the following Key Priority areas: (1) Defence Against Terrorism; (2) Countering other Threats to Security and (3) NATO, Partner and Mediterranean Dialogue Country Priorities. The types of meeting supported are generally "Advanced Study Institutes" and "Advanced Research Workshops". The NATO SPS Series collects together the results of these meetings. The meetings are co-organized by scientists from NATO countries and scientists from NATO's "Partner" or "Mediterranean Dialogue" countries. The observations and recommendations made at the meetings, as well as the contents of the volumes in the Series, reflect those of participants and contributors only; they should not necessarily be regarded as reflecting NATO views or policy.

Advanced Study Institutes (ASI) are high-level tutorial courses intended to convey the latest developments in a subject to an advanced-level audience

Advanced Research Workshops (ARW) are expert meetings where an intense but informal exchange of views at the frontiers of a subject aims at identifying directions for future action

Following a transformation of the programme in 2006 the Series has been re-named and re-organised. Recent volumes on topics not related to security, which result from meetings supported under the programme earlier, may be found in the NATO Science Series.

The Series is published by IOS Press, Amsterdam, and Springer, Dordrecht, in conjunction with the NATO Public Diplomacy Division.

Sub-Series

A.	Chemistry and Biology	Springer
B.	Physics and Biophysics	Springer
C.	Environmental Security	Springer
D.	Information and Communication Security	IOS Press
E.	Human and Societal Dynamics	IOS Press

http://www.nato.int/science
http://www.springer.com
http://www.iospress.nl

Series B: Physics and Biophysics

Evolution from Cellular to Social Scales

edited by

Arne T. Skjeltorp

Institute for Energy Technology,
Kjeller, Norway and Department of Physics,
University of Oslo, Norway

and

Alexander V. Belushkin

Frank Laboratory of Neutron Physics,
Dubna, Russia

Springer

Published in cooperation with NATO Public Diplomacy Division

Proceedings of the NATO Advanced Study Institute on
Evolution from Cellular to Social Scales
Geilo, Norway
10–20 April 2007

Library of Congress Control Number: 2008931060

ISBN 978-1-4020-8760-8 (PB)
ISBN 978-1-4020-8759-2 (HB)
ISBN 978-1-4020-8761-5 (e-book)

Published by Springer,
P.O. Box 17, 3300 AA Dordrecht, The Netherlands.

www.springer.com

Printed on acid-free paper

Preface

This volume comprises the proceedings of a NATO Advanced Study Institute (ASI) held at Geilo, Norway, 10–20 April 2007, the nineteenth ASI in a series held every two years since 1971. The objective of this ASI was to bring together researchers with various interests and background including theoretical physicists, soft condensed matter experimentalists, biological physicists, molecular biologists and social scientists to identify and discuss areas where synergism between modern physics, biology and social sciences may be most fruitfully applied to the study of various aspects of evolution ranging from cellular behaviour to social phenomena including globalization and terrorist networks.

Evolution is a critical challenge for many areas of science, technology and development of society. Emerging areas of science such as "systems biology" and "bio-complexity" are founded on the idea that phenomena need to be understood in the context of highly interactive processes operating at different levels and on different scales. Similarly, there is an increasingly urgent need to understand and predict the evolutionary behavior of highly interacting man-made systems, in areas such as communications and transport, which permeate the modern world. The same applies to the evolution of human networks such as social, political and financial systems, where technology has tended to vastly increase both the complexity and speed of interaction, which is sometimes effectively instantaneous. Better understanding, appreciation and prediction of the behavior of such systems will require the development of tractable methods for addressing evolution. On one hand, this implies deep analysis of particular evolutionary systems. On the other, there is clearly a need to develop general methods of analysis by investigating the similarities and commonalities between evolutionary systems in diverse areas in order to understand, for example, the phenomena in terrorist networks that are common to all networks, or at least to a large class of networks. Many fields of research are confronted with responses to evolutionary pressure in networks. Genetic and metabolic networks have evolved as a result of the interactions of proteins, substrates and genes in a cell. Social networks quantify the interactions between people in the society. Ecological systems are best described as a web of species, and terrorist networks have evolved through political and market pressures. The Internet is a complex web of computers increasingly used by terrorists and increasingly vital to almost all human activities. In many cases the interacting networks manifest regulation and adaptation and so-called emergent properties that are not possessed by any of the individual components. This means that the detailed knowledge of the components is insufficient to describe the whole system.

The starting point in the proposed ASI was a thorough discussion of general evolutionary facts like origin of life and evolution of the genome and clues to evolution through simple systems. This is where physics meets complexity in nature, and where we must begin to learn about complexity if we are to understand it. The next focus was on evolution at different scales and evolution of complex networks in nature and society. Finally, focus was placed on the possible universality of network structures and how this knowledge can be combined to attack the urgent problem of counter threats of security and terrorism.

The scientific content of the school was timely and these proceedings should provide a useful definition of the current status. The Institute brought together many lecturers, students and active researchers in the field from a wide range of countries, both NATO and NATO Partner Countries. The lectures fulfilled the aim of the Study Institute in creating a learning environment and a forum for discussion on the topics stated above. They were supplemented by a few contributed seminars and a large number of poster presentations.

Financial support was principally from the NATO Scientific Affairs Division, but also from the Institute for Energy Technology, the Research Council of Norway and the nationally coordinated research team COMPLEX in Norway.

The editors are most grateful to M.H. Jensen, J. L. McCauley, R. Pynn, N. Stavans and H. Thomas who helped them plan the programme and G. Helgesen for helping with many practical details. Finally, we would like to express our deep gratitude to Trine Løkseth of the Institute for Energy Technology, for all her work and care for all the practical organization before, during, and after the school, including the preparation of these proceedings.

December 2007

Arne T. Skjeltorp
Alexander V. Belushkin

Contents

Organizing Committee and Participants

Organizing Committee

Skjeltorp, Arne T., Director
Institute for Energy Technology
POB 40, N-2027 Kjeller
Norway

Belushkin, Alexander V., Co-Director
Frank Laboratory of Neutron Physics,
Joint Institute for Nuclear
Research, 141980
Dubna, Moscow region
Russia

Helgesen, Geir, Technical Assistant
Institute for Energy Technology
POB 40, N-2027 Kjeller
Norway

Løkseth, Trine, Secretary
Institute for Energy Technology
POB 40, N-2027 Kjeller
Norway

Participants

Agol, Vadim
Institute of Poliomyelitis and Viral
Encephalitides
Moscow
Russia

Agresti, Jeremy Jon
Engineering Sciences Laboratory
40 Oxford Street
Cambridge, MA 02138
USA

Ahlgren, Peter
Niels Bohr Institute
Blegdamsvej 17, 2100 Copenhagen Ø
Denmark

Akbas, Etem
Mersin University – Faculty of Medicine
Department of Medical Biology and
Geneticist
Yenisehir Campus/MERSIN
Turkey

Anisimova, Larysa
Department of Environmental Standards
Institute for Nature Management
Problems and Ecology
National Academy of Sciences of
Ukraine
6 Moskovskaya Street, Dnepropetrovsk
49000
Ukraine

Antunes, Andrei
Av. Joao XXI, 57 ESQ
1000-299 Lisboa
Portugal

Arikan, Mehmet Salýh
Süleyman Demýrel Ünýversýtesý Tip
Fakültesý Hastanesý
Mýkrobýyolojý Laboratuari
Isparta
Turkey

Avlund, Mikkel
Center for Models of Life
Niels Bohr Institute
Blegdamsvej 17, 2100 Copenhagen Oe
Denmark

Babich, Polina
Saint-Petersburg State Polytechnical
University
Kollontay Str. 33, Block 1, Apt. 62
Saint-Petersburg
Russia 193312

Berg, Johannes
Institute for Theoretical Physics
University of Cologne
Germany Technische Universität
Germany

Bergli, Joakim
Fysisk Institutt, University of Oslo
Box 1048 Blindern
0316 Oslo
Norway

Borissova, Daniela
Acad. G. Bonchev Str., Block 2
Sofia – 1113
Bulgaria

Borup, Mia
CMOL
Niels Bohr Institute
Blegdamsvej 17, 2100
Copenhagen Ø
Denmark

Brutovsky, Branislav
Institute of Physics
P. J. Safarik University
Jesenna 5, 04154 Kosice
Slovakia

Carpenter, Holly
Emory University
105 Sycamore Pl., Apt. 603A
Decatur, GA 30030
USA

Cernak, Jozef
University of P.J. Safarik, Department
of Biophysics
Jesenna 5, SK-04000 Kosice
Slovak Republic

Christophorov, Leonid
Bogoliubov Institute for Theoretical
Physics, NAS Ukraine
14 B Metrologichna Str.
Kiev 03143, Ukraine

Dubrovin, Evgeniy
Department of Polymers
Moscow State University
Leninskie Gory, 1/2
Moscow, 119992
Russia

Fossum, Jon Otto
NTNU Department of Physics
Høgskoleringen 5
N-7034 Trondheim
Norway

Giaever, Ivar
Institute of Science, Rensselaer
Polytechnic Institute
Troy, NY 12180
USA

Goksu, A. Yasemin
Istiklal Mh. Fatih Sk. 21/5, Isparta,
32300
Turkey

Hakobyan, Nune
Faculty of Biology, Yerevan State University
Alex Manoogyan Str.1,0025
Yerevan
Armenia

Hauert, Christoph
Program for Evolutionary Dynamics
Harvard University
One Brattle Square
Cambridge, MA
USA

Heiberg-Andersen, Henning
Institute for Energy Technology,
POB 40
NO-2027 Kjeller
Norway

Horvath Denis
Department of Theoretical Physics and Astrophysics
Å afÃ¡rik University
Park Angelinum 9
KoÅ¡ice
Slovakia

Høgh Jensen, Mogens
Niels Bohr Institute, Blegdamsvej 27
DK-2100 KØBENHAVN Ø
Denmark

Jauffred, Liselotte
Niels Bohr Institute
Blegdamsvej 17
2100 Copenhagen
Denmark

Karol, Andrei
Research Center of Spectrometry and Neurography
Leningradskaya 10/16
141980 Dubna
Moscow region
Russia

Kloster, Martin
University of California
San Francisco
USA

Knudsen, Kenneth D.
Institute for Energy Technology,
POB 40
NO-2027 Kjeller
Norway

Krishna, Sandeep
Niels Bohr Institute
Blegdamsvej 17
DK-2100 Copenhagen O
Denmark

Kruchkova, Olga
Bryansk Open Institute of Management and Business
Ul. Orlovskaya 27, kv. 272, 241012
Bryansk
Russia

Lenski, Richard
Michigan State University
East Lansing MI 48824
USA

Lia, Brynjar
Norwegian Defense Research Establishment (FFI)
2007 Kjeller
Norway

Liu, Ying
Niels Bohr Institute
Blegdamsvej 17
2100 Copenhagen
Denmark

Maftuleac, Daniela
Faculty of Mathematics, Moldova State University
Str. Gr. Alexandrescu,17/1, ap.64,
MD-2008
Chisinau
Republic of Moldova

McCauley, Joseph L.
Physics Department
University of Houston, Houston
TX 77204
USA

Meakin, Paul
Laboratory Fellow and Director of the Center for Advanced and Simulation
P.O. Box 1625
Idaho Falls, ID 83415–2211
USA

Medvedeva, Anna
1/12 Leninskie Gory
Department of Developmental Biology
119992 Moscow
Russia

Mengel, Anna Benedicte
Copenhagen University
Priv: Richelieus Alle 11
2900 Hellerup

Micheelsen, Mille Ankerstjerne
The Niels Bohr Institute, University of Copenhagen
Blegdamsvej 17, DK-2100
Copenhagen Ø
Denmark

Milibaeva, Guljamal
Uzbek Academy of Sciences, Heat Physics Department
Laboratory of Perspective Studies
700135
28 Katartal Str., Tashkent
Heat Physics Department
Uzbekistan

Mitarai, Namiko
Niels Bohr Institute
University of Copenhagen
Blegdamsvej 17
2100 Copenhagen Ø

Mohammed, Amjed
Schützenweg 22
26129 Oldenburg
Germany

Mohrdieck, Camilla
Max-Planck Institute for Metal Research
Heisenbergstrasse 3
70569 Stuttgart
Germany

Moxnes, John, F
Norwegian Defence Research Establishment
Kjeller, P.O. Box 25, NO-2007
Norway

Muller, Jiri
Institute for Energy Technology,
POB 40
NO-2027 Kjeller
Norway

Måløy, Knut Jørgen
Fysisk Institutt, University of Oslo
Box 1048 Blindern, 0316 Oslo
Norway

Nepusz, Tamas
Department of Biophysics
Research Institute for Particle and Nuclear Physics
Konkoly-Thege Miklos u. 29–33, 1121 Budapest
Hungary

Neslihan, Turk
Istanbul University
Cerrahpasa Faculty of Medicine
Department of Medical Biology
34098, Cerrahpasa, Istanbul
Turkey

Nikulshin, Vladimir
Ave. Shevchenko, 1
Odessa National Polytechnic University
Odessa, 27044
Ukraine

Olshevskiy, Alexander
Bryansk State Technical University
Bulvar 50-letiya Oktyabrya, 7
Bryansk
Russia, 241035

Ostapchuk, Yuriy
Kyiv National Taras Shevchenko University
Physics Department
Prosp. Glushkova, 2
Kyiv 03022
Ukraine

Ostrovsky, Mikhail
Institute of Bio-Chemical Physics,
Russian Academy of Sciences,
Kosygin St 4
117334 Moscow
Russia

Paiziev, Adkhamjon
Institute of Electronics, Uzbek Academy of Science
F. Khodjaeva Str. 33, Academgorodok
Tashkent 700125
Uzbekistan

Paulsson, Johan Martin
Harvard University
USA

Pokrovsky, Oleg
Main Geophysical Observatory
Karbyshev Str. 7
St. Petersburg, 194021
Russia

Pynn, Roger
Materials Research Lab., UCSB
Santa Barbara
CA 93106–5130
USA

Ramos, Osvanny
Department of Physics
P.O. Box 1048 Blindern
N-0316 Oslo
Norway

Rowat, Amy
Engineering & Sciences Laboratory
40 Oxford Street
Cambridge, MA 02138
USA

Sadoyan, Avetis Abel
Department of Physics
Yerevan State University
Alex Manoogian 1
375025 Yerevan
Armenia

Salomov, Uktam
Heat Physics Department of Uzbek Academy of Sciences
700135
28 Katartal Str., Tashkent
Uzbekistan

Shantsev, Daniel
Department of Physics, University of Oslo
POB 1048 Blindern
NO-0316 Oslo
Norway

Sherrington, David
University of Oxford, Theoretical Physics
1 Keble Rd. Oxford OX1 3NP
UK

Skulachev, Vladimir
Belozersky Institute of Physical-Chemical Biology
Moscow State University
Russia

Smitienko, Olga
117997, Moscow
Kosygina Street, 4
NM Emanuel Institute of Biochemical Physics of the Russian Academy of Sciences
Russia

Sneppen, Kim
Nordita, Blegdamsvej 17, 2100
Copenhagen OE
Denmark

Stavans, Joel
Department of Physics of Complex Systems
Weizmann Institute of Science
PO Box 26, Rehovot 76100
Israel

Steinsvoll, Olav
Institute for Energy Technology
POB 40, NO-2027 Kjeller
Norway

Svenningsen, Sine Lo
Princeton University
USA

Thomas, Harry
Department of Physics, University of Basel
CH-4056 Basel
Switzerland

Trusina, Ala
California Institute for Biomedical Research
University of California at San Francisco
CA 94143
USA

Uhomoibhi, James
Faculty of Engineering
University of Ulster
Shore Road, Newtownabbey
BT37 0QB

Northern Ireland, UK
Vitalie, Eremeev
Institute of Applied Physics
Academy of Sciences of Moldov
Academiei Str. 5, Chisinau
MD-2028
Republic of Moldova

Werner, Maria
Niels Bohr Institute
Blegdamsvej 17, 2100
Copenhagen OE
Denmark

Yazykov, Vladislav
Bryansk State Technical University
bul. 50-letiya Oktyabrya, 7
241035, Russia, Bryansk

Zhukova, Natalia
Dynamics of Geophysical
Fields and Computation
Mathematics
M. Nodia Institute of Geophysics, 1
Alexidze Str, 0193 Tbilisi
Georgia

The Accuracy of Molecular Processes

The Case of Homologous Recombination

Joel Stavans

Abstract Recombination is arguably one of the most fundamental mechanisms driving genetic diversity during evolution. Recombination takes place in one way or another from viruses such as HIV and polio, to bacteria, and finally to man. In both prokaryotes and eukaryotes, homologous recombination is assisted by enzymes, recombinases, that promote the exchange of strands between two segments of DNA, thereby creating new genetic combinations. In bacteria, homologous recombination takes place as a pathway for the repair of DNA lesions and also during horizontal or lateral gene transfer processes, in which cells take in exogenous pieces of DNA. This allows bacteria to evolve rapidly by acquiring large sequences of DNA, a process which would take too long by gene duplications and single mutations. I will survey recent results on the fidelity of homologous recombination as catalyzed by the bacterial recombinase RecA. These results show discrimination up to the level of single base mismatches, during the initial stages of the recombination process. A cascaded kinetic proofreading process is proposed to explain this high discrimination. Kinetic proofreading ideas are also reviewed.

Keywords Homologous, recombination, DNA, RecA, FRET

1 Introduction

A notable feature of bacteria is their promiscuity. It is very common to find them in their natural habitats together with exogenous DNA that they can import, and thus modify their genome. There is a large body of evidence supporting the important

---

Joel Stavans
Department of Physics of Complex Systems
Weizmann Institute of Science
Rehovot 76100
Israel

Arne T. Skjeltorp, Alexander V. Belushkin (eds.), Evolution from Cellular to Social Scales.

role that the process of homologous recombination plays in replacing regions of the genome of a bacterial cell with genomic segments from cells from the same or closely related species [1]. This is allowed by a certain degree of tolerance for differences in sequence between the participating DNA segments, in homologous recombination carried out by the bacterial protein RecA [2–4]. RecA participates both in the recombination repair pathway of DNA lesions, as well as in lateral gene transfer processes. The tolerance of RecA-catalyzed homologous recombination is held in check by the mismatch repair systems (MRS) such as the Mut proteins in *E. coli*. There are situations in which MRS systems either play a minor role, or are down-regulated. An example is the stationary phase of bacteria [5, 6], which happens to be their most common lifestyle. In these situations, the only discrimination against mismatches between an exogenous DNA sequence and the genome is provided by RecA itself. In these lectures I will start by reviewing basic facts about homologous recombination, survey recent results of experiments in which the fidelity of RecA-catalyzed homologous recombination was tested *in vitro*, and finally, discuss a recent model designed to explain how can recombination exhibit discrimination to the level of a single mismatch, during the initial stages of the process.

1.1 Homologous Recombination

In both prokaryotes and eukaryotes, homologous recombination proceeds according to the following steps (Fig. 1). A single-stranded DNA substrate is created (e.g. upon exogenous DNA uptake in *P. pneumoniae* or during the repair of a double strand break in DNA). Next, a recombinase such as RecA in the case of bacteria

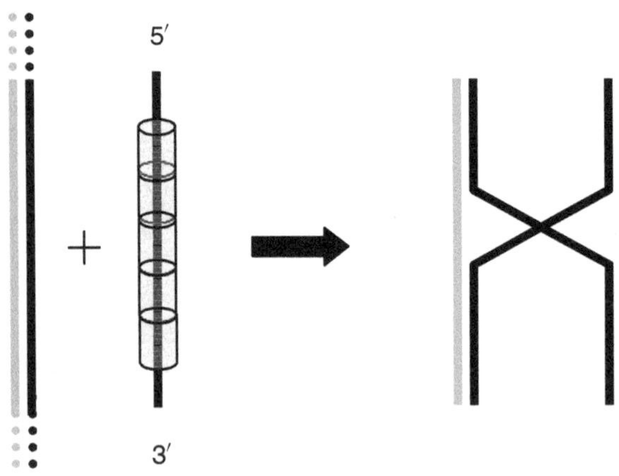

Fig. 1 Strand exchange between double-stranded DNA and a nucleoprotein complex formed by single-stranded DNA on which a recombinase (shaded cylinders) polymerizes. For strand exchange to occur, the degree of homology between the ssDNA and the dsDNA segments must be sufficiently high

polymerizes on this substrate from the 5′ to 3′ direction of the ssDNA. Interestingly the RecA filament may depolymerize, also in the 5′ to 3′ direction, a process that requires ATP hydrolysis. While evidence for polymerization-depolymerization cycles exists, their functional significance is unclear.

Next a process of homology search ensues, whereby the RecA-ssDNA nucleoprotein complex tests the genome for tracts of high enough homology with the ssDNA. The RecA-ssDNA aligns with the dsDNA, forming a *synaptic complex*. It is within the synaptic complex that homology is thought to be probed. When an homologous tract is found, RecA-catalyzed strand exchange ensues. After strand exchange, DNA replication takes care of filling up the rest of single-stranded regions and the cruciform structures are then resolved, leaving two intact pieces of dsDNA. The search for homology as well as strand exchange functions are so fundamental that they have been preserved during evolution: in higher organisms including man, they are carried out by the Rad51 and Dmc1 proteins [7], which share a high degree of homology with RecA. Rad51 participates in DNA repair, while Dmc1 is involved exclusively in chromosome pairing.

As we may expect, the process of homologous recombination is not error free. Indeed, RecA-assisted strand exchange can take place, albeit with a smaller efficiency, if the exchanged strands differ by a number of mismatches, insertions and deletions [4, 8]. It was widely held that a percentage of mismatches above 10% was necessary to see their effect on the efficiency of recombination [9], and that recombination was more error-prone than simple hybridization [10]. It thus came as a surprise when recent experiments provided evidence for *single* mismatch discrimination, during the initial stages of recombination [8], and for high sensitivity of the efficiency of recombination to the location and distribution of mismatches. This discrimination is most sensitive when the mismatches are located within the first ∼30 bases of the invading 3′ strand, providing for the first time an explanation in molecular terms for the existence of a minimal efficient processing DNA segment length (MEPS), over which homology is probed, and below which recombination becomes inefficient *in vivo* [11]. Importantly, from the Physics point of view, this discrimination is well above that expected from differences in binding energies between the replaced strand on the dsDNA and the invading strand. Hence *equilibrium* Statistical Mechanics and Thermodynamics cannot account for the high fidelity of recombination in its initial stages.

2 Experimental Study of RecA-Induced Homologous Recombination

In order to address the problem of the fidelity of RecA-induced homologous recombination and shed light on the problem of target location, *in vitro* homologous recombination experiments with short DNA oligomers were carried out, using fluorescence resonant energy transfer techniques (FRET). The technique is first described briefly, and then a short survey of the main results follows.

2.1 Fluorescence Resonant Energy Transfer (FRET)

FRET is based on labeling the participating molecular species with two fluorescent groups: a donor and an acceptor. A reaction volume is then illuminated with light with a wavelength within the excitation spectrum of the donor, and as outside as possible from the excitation spectrum of the acceptor. The donor then emits fluorescence. However, if the acceptor happens to be sufficiently near, then there can be a *radiationless* transfer of energy from donor to the acceptor, which then emits with fluorescence within its emission spectrum. This only happens essentially when both fluorophores are within $\sim$100 Å. For distances r beyond this limit FRET is essentially undetectable due to the strong $1/r^6$ dependence of the underlying dipole–dipole interaction between the two fluorophores.

One possible way to implement the technique in homologous recombination is to label the strands of a dsDNA one with donor and the other with acceptor, at the same end of the dsDNA. When ssDNA covered with RecA is then added, strand exchange separates donor from acceptor, with a consequent decay in the FRET signal. Alternatively, one can also label the dsDNA, say with a donor and the ssDNA with the acceptor, and follow the increase in the FRET signal. In our experiments we used Tamra as a donor and Cy5 as an acceptor. Our oligomers were about 50 bp in length, and enough ATP was added so that its concentration is not depleted by the reaction. For more details about the experiment one can consult reference [8].

2.2 Results

I show in Fig. 2B the fraction of strand exchange as function of RecA concentration for different invading ssDNA sequences shown in Fig. 2A. The position of the mismatches is shown with red letters. Notice that one single mismatch near the invading $3'$ reduces the efficiency of strand exchange considerably, and 2–3 mismatches essentially reduce it to that near full heterology. These results were at first surprising: how can one find consistency with previous experiments in which RecA-catalyzed homologous recombination was observed to be much more tolerant to heterologies [9, 10]?

To answer this question, we carried out experiments in which we tested the behavior of the strand exchange efficiency as we moved a single mismatch from the $3'$ to the $5'$ end of the incoming strand. The results are shown in Fig. 2C, for two completely different target dsDNA sequences. The reduction in the efficiency of strand exchange due to the presence of a mismatch progressively goes down, as the distance of the mismatch to the $3'$ end increases. A mismatch near the $5'$ end has barely an effect. This experiment suggests that (i) the strand exchange process is directional, (ii) starts by the invasion of the $3'$ end of the incoming strand, and (iii) once the strand exchange process is well advanced, i.e. far from the $3'$ end, the newly-formed dsDNA stabilizes the synaptic complex, making it harder to dissociate, so that further strand exchange can continue.

To mimic the search for homology in a cell, experiments in which RecA-catalyzed strand exchange between dsDNA and a fully homologous ssDNA were

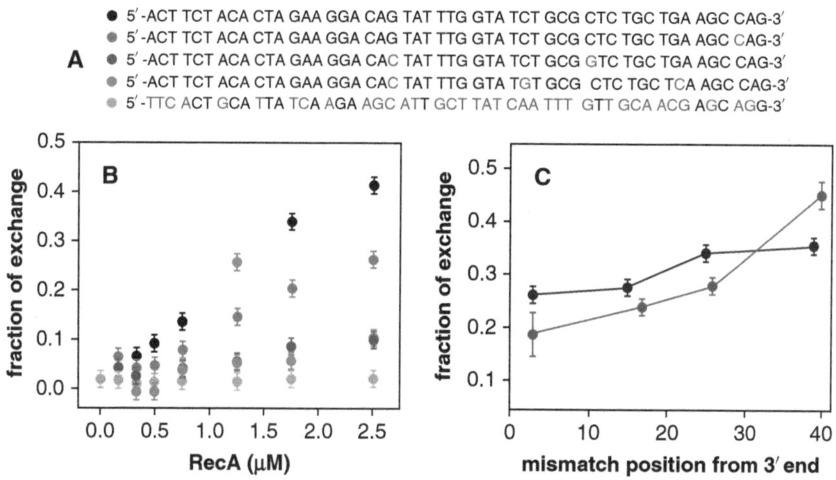

Fig. 2 Effect of mismatches on strand exchange. (A) Sequences. (B) Fraction of exchange as a function of RecA concentration, for the sequences in (A). (C) The effects on strand exchange of moving a single mismatch from the 3′ to the 5′ end on the invading ssDNA strand

carried out in the presence of competitor dsDNA. The competitor dsDNA, present in an excess of about ten times the concentration of the duplex undergoing strand exchange, bore full homology with the latter, but only up to x base pairs from one end. In the experiment the delay of the strand exchange reaction due to the competitor was measured, for different values of x. A plot of the rise time of the FRET signal as strand exchange proceeds is shown in Fig. 3. The rise time of the reaction in the absence of competitor (white circle) is also shown. Two features are noteworthy:

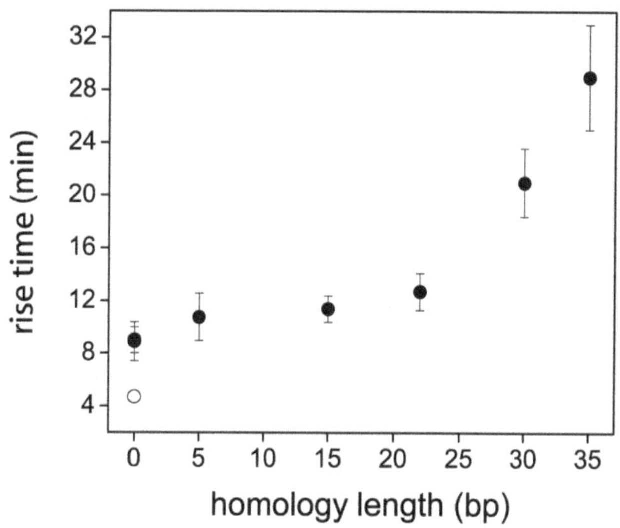

Fig. 3 Rise time of the fraction of strand exchange between fully homologous dsDNA and ssDNA, in the presence of competitor duplexes homologous to the dsDNA up to base pair x from one of its ends. The competitor's concentration is 10x that of the dsDNA

first, there is a delay of about 5 min in the reaction, when fully heterologous competitor is added. This implies that in spite of the lack of homology, the competitor is able to sequester for a significant time the ssDNA-RecA complex, while the degree of homology is read in the synapse. Furthermore, a rough estimate of the synapse lifetime yields ~10 sec, a lifetime which is huge compared to typical molecular timescales [8].

Second, the rise time changes little while $x \leq 23$, but then increases rapidly above this value. In other words, the competitor's effectiveness starts increasing when the duplex incorporating the invading strand becomes sufficiently large to stabilize the synapse against dissociation. These experiments provide a molecular basis for the *in vivo* observation of a MEPS of around 30 bases [12].

3 A Kinetic Proofreading Cascade Scheme for Homologous Recombination

Consider right C and wrong D substrates on which an enzyme E works, giving either right or wrong products in a Michaelis-Menten scheme.

$$C + E \underset{k_C}{\overset{k'_C}{\rightleftharpoons}} CE \xrightarrow{W} \text{correct product}$$

$$D + E \underset{k_D}{\overset{k'_D}{\rightleftharpoons}} DE \xrightarrow{W} \text{incorrect product}$$

We assume for simplicity that the concentrations of C and D to be the same and that the rate W of the last step is small, so that the concentrations of the intermediates CE, DE achieve near equilibrium values relative to the first step. The rate W being the same in both reactions, the ratio of concentrations of right and wrong products will be given by the Bolzmann factor f_o of the free energy difference:

$$f_o = \exp(-\Delta G_{CD}/kT) = (W + k_c)/(W + k_D)$$

In other words, in this simple scheme, the discrimination is determined by f_0. The question is, can we do better? The answer is yes, as Ninio and Hopfield have independently shown [13, 14]. To see this, let us first augment the simple scheme with an additional intermediate step, from which the substrate can be discarded by a new pathway with rate ℓ:

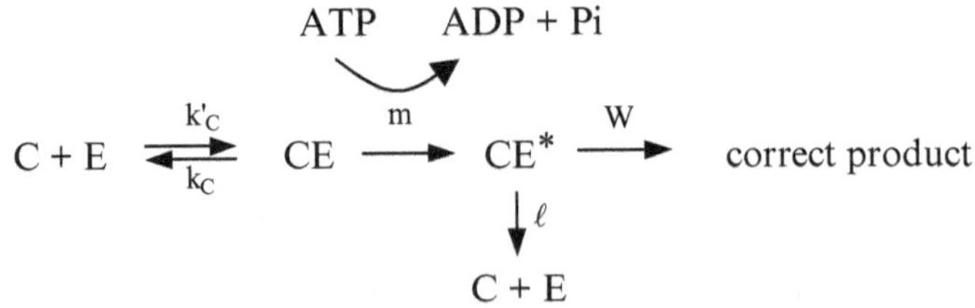

An analogous reaction for D is assumed. CE^* is assumed to be a high energy state reached effectively only from CE by using energy, but not from $C+E$. Hence the reactions with rates m and ℓ are essentially irreversible. Now, if $k_C \ll m$ then the first reaction can be assumed to be near equilibrium, and so the ratio of concentrations of CE and DE is $\sim f_o$. Under these conditions one can obtain discrimination approaching f_o^2: the new intermediate CE^* decays faster than DE^* by a factor f_o and its formation is already biased by an additional factor f_o from the first step. One can of course add more intermediate steps in a cascade, say n steps, couple each to ATP hydrolysis, and achieve a discrimination approaching $\sim f_o^n$. This is the gist of the idea. Had we not made the new intermediate CE^* a high energy one, detailed balance considerations show that the addition of another step along the reaction does not bring about any improvement in discrimination above f_o.

To account for the extraordinary accuracy of the initial stages of recombination catalyzed by RecA, Sagi et al. [8] proposed a cascaded scheme of *kinetic proofreading* stages, in which the invasion of the new ssDNA strand proceeds sequentially and directionally from stage i to stage $i+1$ at a rate α_0, and homology is checked at each stage. Advance is coupled to an energy source (ATP hydrolysis) in order to drive the system out of equilibrium. Furthermore, extra pathways (with off rate β_i) are provided, in order to allow for synaptic complex dissociation. It is further assumed in our model that each stage is more stable than the previous one, by virtue of the longer extent of the duplex formed by the strand exchange process ($\beta_i > \beta_{I+1}$) as schematized below:

One can write master equations appropriate to this scheme, and calculate the effect of mismatches on the efficiency of strand exchange. The model, while appearing consistent with the experimental results, remains to be confirmed experimentally.

4 Discussion

Two characteristics of the strand exchange process are critical for achieving single mismatch discrimination during the initial stages of homologous recombination: directionality and the use of an energy source to drive reactions out of equilibrium. These characteristics motivated the use of kinetic proofreading ideas to account for the observed discrimination. The model ties nicely the fact that for length scales below the that of the MEPS discrimination is high, whereas above this length scale, discrimination is highly reduced and recombination can accept large heterologies. In spite of these nice features, further experiments are needed to provide more evidence either in favor or refuting this model.

References

1. Spratt, B.G., Hanage, W.P. and Feil, E.J. (2001) The relative contributions of recombination and point mutation to the diversification of bacterial clones. *Curr. Opin. Microbiol.*, **4**, 602–606.

2. Bianchi, M.E. and Radding, C.M. (1983) Insertions, deletions and mismatches in heteroduplex DNA made by recA protein. *Cell*, **35**, 511–520.
3. Morel, P., Stasiak, A., Ehrlich, S.D. and Cassuto, E. (1994) Effect of length and location of heterologous sequences on RecA-mediated strand exchange. *J. Biol. Chem.*, **269**, 19830–19835.
4. Bucka, A. and Stasiak, A. (2001) RecA-mediated strand exchange traverses substitutional heterologies more easily than deletions or insertions. *Nucleic Acids Res.*, **29**, 2464–2470.
5. Feng, G., Tsui, H.C.T. and Winkler, M.E. (1996) Depletion of the cellular amounts of the MutS and MutH methyl-directed mismatch repair proteins in stationary-phase Escherichia coli K-12 cells. *J. Bacteriol.*, **178**, 2388–2396.
6. Bregeon, D., Matic, I., Radman, M. and Taddei, F. (1999) Inefficient mismatch repair: genetic defects and down regulation. *J. Genet.*, **78**, 21–28.
7. Masson, J.Y. and West, S.C. (2001) The Rad51 and Dmc1 recombinases: a non-identical twin relationship. *Trends Biochem. Sci.*, **26**, 131–136.
8. Sagi, D., Tlusty, T. and Stavans, J. (2006) High fidelity of RecA-catalyzed recombination: a watchdog of genetic diversity. *Nucleic Acids Res.*, **34**, 5021–5031.
9. Bazemore, L.R., FoltaStogniew, E., Takahashi, M. and Radding, C.M. (1997) RecA tests homology at both pairing and strand exchange. *Proc. Natl. Acad. Sci. USA*, **94**, 11863–11868.
10. Malkov, V.A., Sastry, L. and Camerini-Otero, R.D. (1997) RecA protein assisted selection reveals a low fidelity of recognition of homology in a duplex DNA by an oligonucleotide. *J. Mol. Biol.*, **271**, 168–177.
11. Majewski, J. and Cohan, F.M. (1999) DNA sequence similarity requirements for interspecific recombination in bacillus. *Genetics*, **153**, 1525–1533.
12. Shen, P. and Huang, H.V. (1986) Homologous recombination in *Escherichia coli*: dependence on substrate length and homology. *Genetics*, **112**, 441–457.
13. Ninio, J. (1975) Kinetic amplification of enzyme discrimination, *Biochimie*, **57**, 587.
14. Hopfield, J.J. (1974) Kinetic Proofreading: A New Mechanism for Reducing Errors in Biosnythetic Processes Requiring High Specificity. *Proc. Natl. Acad. Sci. USA*, **71**(10), 4135–4139.

Experimental Evolution: Bugs and Bytes

Richard E. Lenski

Abstract Comparative studies of living organisms and their genomes, along with the fossil record, provide evidence for the history of life on Earth and the evolutionary processes that have shaped living systems. However, it can be difficult or impossible to resolve certain issues or test particular hypotheses given the inherent limitations in such observational data. The field of experimental evolution seeks to overcome these limitations by using model systems to test particular hypotheses and explore complex issues by means of rigorously replicated and controlled experiments. Model organisms typically have rapid generations, are easy to grow, can be stored in suspended animation, and are amenable to precise environmental and genetic manipulations.

In my group, we employ two experimental systems for studying evolution, 'bugs' and 'bytes'. The former refers to microorganisms including the bacterium *Escherichia coli* (among others that we have studied). The latter refers to digital organisms – by which I mean computer programs that can self-replicate, mutate, and compete with one another – that evolve by the same fundamental processes that underlie evolution in nature.

The reference list below begins with two synthetic review articles; the first covers experimental evolution with bacteria and other microorganisms [1], while the second review explores experiments on digital organisms [2]. The next three papers report original research in these two systems; these should allow one to delve more deeply into the field of experimental evolution, in general, and the two experimental systems that we study in my group, in particular [3–6]. The final paper explores the future of evolution [7]. It identifies four areas for future investigation that, in my view, offer tremendous excitement: (i) the molecular genetic underpinnings of adaptation, (ii) the evolution of evolvability, (iii) artificial life as a model system for understanding evolution, and, most provocatively, (iv) artificial life as a future component of our evolving world.

Richard E. Lenski
Michigan State University
East Lansing, MI 48824, USA

Arne T. Skjeltorp, Alexander V. Belushkin (eds.), Evolution from Cellular to Social Scales.

Keywords Evolution experiments with microorganisms, dynamics of adaptation and diversification, parallel molecular evolution, evolutionary biology.

References

1. Elena, S. F., and R. E. Lenski. 2003. Evolution experiments with microorganisms: the dynamics and genetic bases of adaptation. *Nature Rev Genet* **4**:457–469.
2. Adami, C. 2006. Digital genetics: unravelling the genetic basis of evolution. *Nature Rev Genet* **7**:109–118.
3. Lenski, R. E., and M. Travisano. 1994. Dynamics of adaptation and diversification: a 10,000-generation experiment with bacterial populations. *Proc Natl Acad Sci, USA* **91**:6808–6814.
4. Woods, R., D. Schneider, C. L. Winkworth, M. A. Riley, and R. E. Lenski. 2006. Tests of parallel molecular evolution in a long-term experiment with *Escherichia coli. Proc Natl Acad Sci, USA* **103**:9107–9112.
5. Blount, Z. D., C. Z. Borland, and R. E. Lenski. 2008. Historical contingency and the evolution of a key innovation in an experimental population of *Escherichia coli. Proc Natl Acad Sci, USA* **105**:7899–7906.
6. Lenski, R. E., C. Ofria, R. T. Pennock, and C. Adami. 2003. The evolutionary origin of complex features. *Nature* **423**:139–144.
7. Lenski, R. E. 2004. The future of evolutionary biology. *Ludus Vitalis* **12**:67–89 [Available at www.msu.edu/~lenski/].

Evolutionary Dynamics

Christoph Hauert

Abstract Evolutionary dynamics in finite populations reflects a balance between Darwinian selection and random drift. For a long time population structures were assumed to leave this balance unaffected provided that the mutants and residents have fixed fitness values. This result indeed holds for a certain (large) class of population structures or graphs. However, other structures can tilt the balance to the extend that either selection is eliminated and drift rules or drift is eliminated and only selection matters.

In nature, however, fitness is generally affected by interactions with other members of the population. This is of particular importance for the evolution of cooperation. The essence of this evolutionary conundrum is captured by social dilemmas: cooperators provide a benefit to the group at some cost to themselves, whereas defectors attempt to exploit the group by reaping the benefits without bearing the costs of cooperation. The most prominent game theoretical models to study this problem are the prisoner's dilemma and the snowdrift game. In the prisoner's dilemma, cooperators are doomed if interactions occur randomly. In structured populations, individuals interact only with their neighbors and cooperators may thrive by aggregating in clusters and thereby reducing exploitation by defectors. In finite populations, a surprisingly simple rule determines whether evolution favors cooperation: $b > c\ k$ that is, if the benefits b exceed k-times the costs c of cooperation, where k is the (average) number of neighbors. The spatial prisoner's dilemma has lead to

Christoph Hauert (✉)
Program for Evolutionary Dynamics
Harvard University
One Brattle Square
Cambridge MA 02138, USA

Present address:
Department of Mathematics
University of British Columbia
1984 Mathematics Road
Vancouver B.C. Canada V6T 1Z2
christoph.hauert@math.ubc.ca

the general belief that spatial structure is beneficial for cooperation. Interestingly, however, this no longer holds when relaxing the social dilemma and considering the snowdrift game. Due to the less stringent conditions, cooperators persist in populations with random interactions but spatial structure tends to be deleterious and may even eliminate cooperation altogether.

In many biological situations it seems more appropriate to assume a continuous range of cooperative investment levels instead of restricting the analysis to two a priori fixed strategic types. In the continuous prisoner's dilemma cooperative investments gradually decrease and defection dominates just as in the traditional prisoner's dilemma. In contrast, the continuous snowdrift game exhibits rich dynamics but most importantly provides an intriguing natural explanation for phenotypic diversification and the evolutionary origin of cooperators and defectors. Thus, selection may not always favor equal contributions but rather promote states where two distinct types co-exist – those that fully cooperate and those that exploit. In the context of human societies and cultural evolution this could be termed the *Tragedy of the Commune* because differences in contributions to a communal enterprise have significant potential for escalating conflicts on the basis of accepted notions of fairness.

Keywords Evolutionary game theory, evolutionary graph theory, social dilemmas, prisoner's dilemma, snowdrift game, structured populations, continuous games, evolutionary branching

1 Modeling Evolution

Evolutionary dynamics acts on populations – neither genes, nor cells, nor individuals evolve, only populations evolve. Conversely, Darwinian selection does not act on populations but on genes, cells and individuals. Selection reflects the fact that the genes or behavioral patterns of individuals with a higher fitness have a higher chance to be passed to subsequent generations through biological reproduction or cultural imitation. For an excellent introduction into evolutionary dynamics see Nowak (2006). In a nutshell, the evolutionary process is determined by:

Selection: Individuals with a fitness that exceeds the average fitness in the population have a higher propensity to pass their genetic or cultural traits to progeny in subsequent generations and these traits are likely to increase in abundance. Similarly, traits that lower the fitness of an individual have small chances to be passed to the next generation, decrease in abundance and eventually disappear.

Variation: Mutations and genetic recombination as well as spontaneous alterations and erroneous imitations of behavioral patterns generate fitness differences among members of the population. Selection acts on these differences and amplifies them over time.

Random drift: In finite populations the transmission of traits through reproduction or imitation is generally a stochastic process. The success of a trait is proportional to the fitness of its carriers but a high fitness does not provides any guarantee for success. With a small probability even the fittest member of the population may not get a chance to pass its trait to the next generation. Similarly, traits of even the least fit individual may get an odd chance to proliferate and persist in the population through random drift. Random drift counteracts selection and becomes increasingly important in smaller populations or for decreasing fitness differences within a population.

Selection, variation and random drift represent the makeup of Darwin's survival of the fittest. It is important, however, to recognize that evolution represents a myopic optimization process. Even though selection always favours individuals with higher fitness, this does not necessarily imply that the average fitness of the population increases. Quite on the contrary, it is often the case that evolution favours traits that reduce the overall fitness of the population. This fundamental problem will become most apparent when addressing the conundrum of the evolution of cooperation in Sect. 3. The reason for such outcomes lies in the fact that the fitness of an individual is not entirely genetically or culturally predetermined but instead depends on interactions with other members of the population and thus depends on the current abundance and possibly the distribution of traits in the population.

1.1 The Replicator Dynamics

Let us start with the simplest possible evolutionary scenario: an infinitely large population consists of two types, the residents and the mutants. Individuals do not interact and thus the fitness of both types is fixed and independent of their relative abundances. Residents have a normalized fitness of $f_r = 1$ and mutants have a fitness $f_m = r$. If x denotes the fraction of mutants in the population then the evolutionary fate of the mutants is given by the replicator equation (Hofbauer and Sigmund, 1998):

$$\dot{x} = x(f_m - \bar{f}), \tag{1}$$

where $\bar{f} = xf_m + (1-x)f_r$ denotes the average population payoff. This simply means that if $f_m > \bar{f}$ then the mutants increase in abundance. Here this condition reduces to $r > 1$. Thus, if the mutant has a higher fitness than the resident, the fraction of mutants keeps increasing until eventually the mutant displaces the resident and thus becomes the new resident. Conversely, if $r < 1$ the mutant is bound to disappear irrespective of its initial frequency. The evolutionary process is deterministic – if $r > 1$ mutants take over with certainty and disappear with certainty for $r < 1$.

1.2 The Moran Process

Evolution in finite populations is stochastic and can be modeled using the Moran process (Moran, 1962): in every time step an individual is randomly selected for reproduction with a probability proportional to its fitness and produces a single clonal offspring that replaces a randomly selected member of the population (see Fig. 1). The total population size N remains constant, i.e. the Moran process assumes that N represents the carrying capacity of the population and neglects fluctuations in population size. All individuals have the same average lifespan but fitter individuals tend to have a higher reproductive output. This represents a specific balance between selection and random drift: fitter individuals have higher chances – but no guarantee – of reproduction, whereas less fit individuals are likely – but again, no guarantee – to be eliminated.

Returning to our simplest possible evolutionary scenario, repeatedly applying the above updating procedure determines the evolutionary fate of residents and mutants. In the absence of mutations the Moran process ultimately results in a homogeneous population with all residents or all mutants because, irrespective of the initial configuration, eventually all members of the population will have a single common ancestor. Both homogeneous states are absorbing and represent an evolutionary end state. The remaining type is said to have reached fixation. We can now determine the probability that mutants (or residents) fixate for a particular initial configuration.

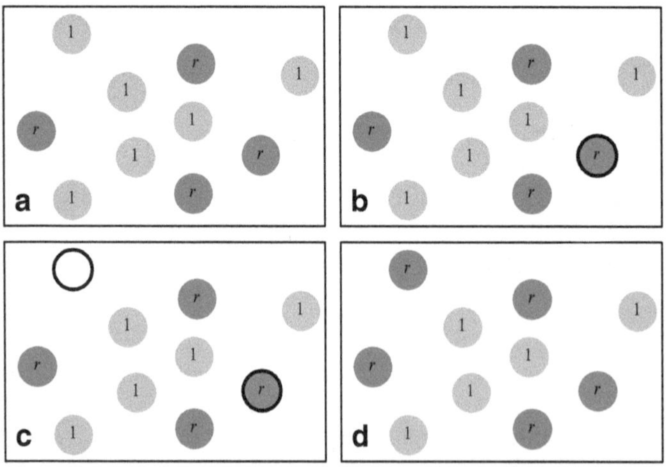

Fig. 1 The Moran process describes the stochastic evolution of a finite population of constant size. **a** The population consists of a mixture of residents (light grey, fitness 1) and mutants (dark grey, fitness r). **b** A focal individual is randomly selected for reproduction with a probability proportional to its fitness. **c** A randomly selected individual (independent of fitness) is removed. **d** The vacancy is replaced by a clonal offspring of the focal individual

In unstructured populations the state of the population is fully determined by the number of mutants present. The number of mutants i changes at most by ± 1 in every time step of the Moran process. With probability T^+ the number of mutants increases from i to $i+1$, with probability T^- it decreases to $i-1$ and with probability $1 - T^+ - T^-$ the number of mutants remains unchanged.

$$T^+ = \frac{i \cdot r}{i \cdot r + (N-i)} \cdot \frac{N-i}{N} \tag{2a}$$

$$T^- = \frac{N-i}{i \cdot r + (N-i)} \cdot \frac{i}{N} \tag{2b}$$

The first factor of T^+ (T^-) indicates the probability that a mutant (resident) is chosen for reproduction and the second factor denotes the probability that the offspring replaces a resident (mutant). Note that the ratio of the transition probabilities $T^+/T^- = r$ is independent of the number of mutants in the population. This leads to a simple recursive formula for the fixation probability $\rho(i)$ of the mutant in a population with i mutants:

$$\rho(i) = \frac{r}{1+r}\rho(i-1) + \frac{1}{1+r}\rho(i+1). \tag{3}$$

Thus, the dynamics corresponds to a biased random walk with absorbing boundaries. Equation (3) admits two solutions $\rho = 1$ and $\rho = 1/r^i$. The absorbing boundaries additionally require $\rho(0) = 0$ and $\rho(N) = 1$. For $r \neq 1$, the fixation probability of a single mutant ρ_1 then becomes

$$\rho_1 = \frac{1 - \frac{1}{r}}{1 - \frac{1}{r^N}}. \tag{4}$$

Assuming that mutations are rare events ρ_1 is of particular interest. It is easy to see that a neutral mutant ($r = 1$) has a fixation probability of $\rho_1 = 1/N$: eventually the entire population will have a single common ancestor but in terms of fitness mutants and residents are indistinguishable and so every member of the population has equal chances to be the chosen one. Evolution is said to favor a mutant if the fixation probability of the mutant exceeds the fixation probability of a neutral mutant, $\rho_1 > 1/N$.

In contrast to the replicator dynamics, Eq. (4) shows that evolution favors mutants for $r > 1$ but it is no longer a guarantee to reach fixation and similarly, for $r < 1$ fixation is less likely but again no guarantee that the mutation disappears. The balance between selection and random drift depends on the population size N. For small N random drift dominates whereas for large N selection becomes more important and in the limit $N \to \infty$ the deterministic replicator dynamics is recovered (Traulsen et al., 2005).

2 Evolutionary Graph Theory

Using evolutionary graph theory (Lieberman et al., 2005) we can investigate the effects of population structures on the evolutionary process. In structured populations, individuals occupy the nodes of a network or graph of size N. The links between the nodes define the neighborhood of each individual. The graph can have any structure – for example, square lattices describe spatially extended systems or small-world networks (Watts and Strogatz, 1998) model social structures – and links between nodes may have different strengths. A fully connected graph, where each node is equally linked to every other node, is equivalent to an unstructured population. Mathematically, the structure of the graph is determined by the adjacency matrix $W = [w_{ij}]$ where w_{ij} denotes the strength of the link between nodes i and j. If $w_{ij} = 0$ and $w_{ji} = 0$ then the two nodes i and j are not connected.

The original Moran process is easily adapted to model evolution on arbitrary graphs (see Fig. 2): in every time step a focal individual i is randomly selected for reproduction with a probability proportional to its fitness and produces a single clonal offspring that replaces a random *neighbor* j, which is selected with a probability proportional to w_{ij}.

In structured populations the essential difference is that the offspring replaces a neighbor of the focal individual instead of a random member of the population. An additional minor difference is that the offspring cannot replace the focal individual but, in principle, this can be implemented by adding loops to each node.

Provided that the graph is connected – each node is connected to every other node through a series of links – the system has the same two absorbing states with

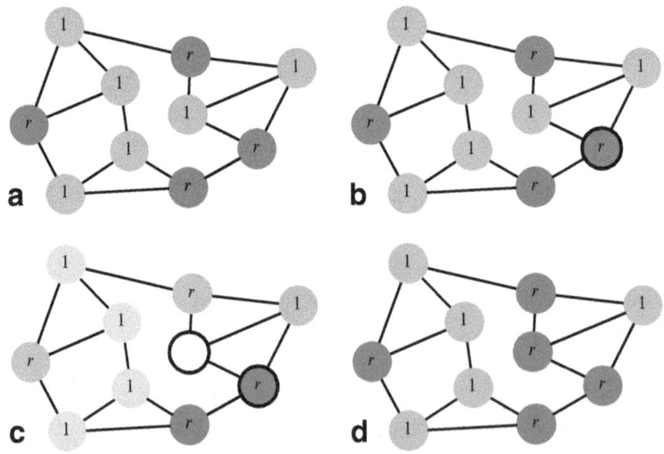

Fig. 2 Moran process on graphs – stochastic evolution in a finite structured population of size N. All individuals occupy the nodes of a graph or network where links between nodes determine each individual's neighborhood. **a** The population consists of a mixture of residents (light grey, fitness 1) and mutants (dark grey, fitness r). **b** A focal individual i is randomly selected for reproduction with a probability proportional to its fitness. **c** A random selected neighbor j of the focal individual is removed. **d** The vacancy is filled by clonal offspring of the focal individual

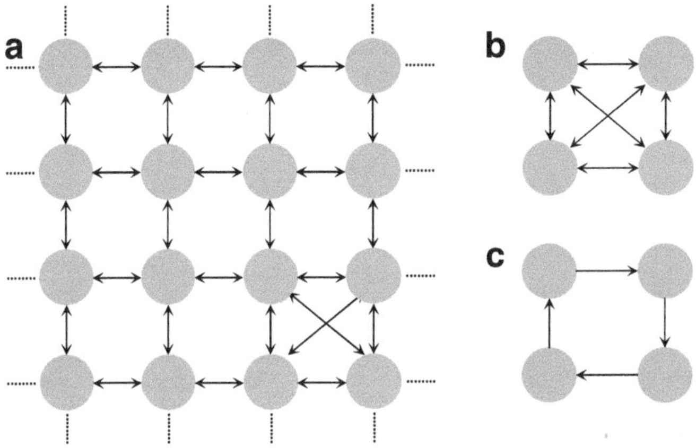

Fig. 3 Sample graphs illustrating different simple population structures. What is the fixation probability of a single mutant on each of these graphs? Is it more or less likely than in an unstructured population? **a** Regular lattice where each individual has four neighbors. **b** Fully connected graph. **c** Cyclic structure with directed links – replacement does not need to go both ways

all residents or all mutants. In the following we always assume connected graphs because otherwise the dynamics must be considered for each subgraph individually. Thus, we can again ask the question about the fixation probability of a single mutant in a structured resident population. In other words, how does the limited dispersal of offspring in structured populations affect the fixation probability of mutants? Some simple sample graphs are shown in Fig. 3. Is fixation easier or harder on these graphs? – or is it independent of the population structure? In fact, the celebrated results by Maruyama (1970) and Slatkin (1981) indicate that the latter is the case. And indeed, all graphs in Fig. 3 leave the fixation probability unchanged and ρ_1 is the same as in unstructured populations. However, this is not true in general but it does hold for a broad class of graphs.

2.1 Circulation Theorem

In order to characterize the class of graphs that lead to the same fixation probability ρ_1 as the original Moran process in unstructured populations, let us introduce the flux through each node. The sum of the weights of the incoming links $f_i^{\text{in}} = \Sigma_j w_{ji}$ denotes the flux entering node i. f_i^{in} relates to a temperature because it indicates the rate at which the occupant of node i gets replaced. 'Hot' nodes are frequently replaced and 'cold' nodes only rarely change their type. In analogy, the sum of the weights of the outgoing links $f_i^{\text{out}} = \Sigma_j w_{ij}$ denotes the flux leaving node i. f_i^{out} determines the impact of node i on its neighborhood. The matrix W is a circulation if $f_i^{\text{in}} = f_i^{\text{out}}$ holds for all nodes.

Circulation Theorem: The Moran process on a graph results in the same fixation probability ρ_1 of a single mutant as in an unstructured population if and only if the graph is a circulation.

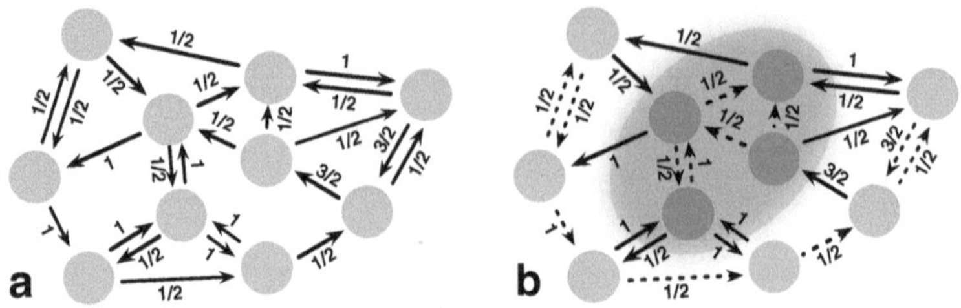

Fig. 4 More general graph where mutants have the same fixation probability than in an unstructured population. The graph is a circulation with an asymmetric adjacency matrix W. The arrows indicate the direction and the labels the strength of the links. **a** The flux entering each node is balanced by the flux leaving the node, $f_i^{\text{in}} = f_i^{\text{out}}$. This is the definition of a circulation. **b** Mutants (dark grey) occupy a connected subset of the graph (shaded area) and are surrounded by residents (light grey). Note that for the mutant subset holds the same flux balance as for every node: the sum of the weights of incoming links (connecting residents to mutants) is the same as the sum of the weights of outgoing links (connecting mutants to residents). Evolutionary dynamics changes the composition of the population only if replacements occur along one of the solid arrows (connecting the mutant subset and the residents) but not along the dashed arrows. However, the flux balance of the mutant subset is not affected by adding or removing mutants. Because of this invariance mutants have the same fixation probability on a circulation graph as in an unstructured population (see text)

In particular, the circulation theorem holds if W is symmetric ($w_{ij} = w_{ji}$), which holds for all graphs shown in Fig. 3, or if W is isothermal, i.e. all nodes have the same temperature ($f_i^{\text{in}} = T \ \forall i$). Another special case of the circulation theorem applies if W is doubly stochastic, i.e. if all rows and columns of W sum up to one. A more general circulation graph where W is not symmetric is depicted in Fig. 4a. A detailed proof of the circulation theorem is provided in Lieberman et al. (2005). Here we provide an intuitive illustration of the circulation conditions and its consequences.

At any point in time during the invasion process of mutants on a circulation graph, it is possible to identify connected subsets of nodes on the graph that are occupied by mutants such that all adjacent nodes of each subset are occupied by residents. Obviously, the state of the population changes only if a replacement occurs along one of the links connecting residents and mutants. Figure 4b shows a general circulation graph with one connected subset (shaded area) of mutants. Multiple such subsets may exist and, in fact, the evolutionary process may split large connected subsets of mutants into two smaller ones or may merge two previously unconnected subsets into one larger subset. For each subset, the circulation theorem requires that the sum of the weights of links pointing out of the subset (connecting a mutant node to an adjacent resident node) equals the sum of the weights of links pointing into the subset (connecting an adjacent resident with a mutant node within the subset). Since the influx is balanced by the outflux, $f_i^{\text{in}} = f_i^{\text{out}}$, for all nodes, increasing the mutant subset by replacing an adjacent resident node with a mutant or decreasing the subset by replacing a mutant with a resident, does not affect the flux balance of the subset. For the same reason, the balance remains unchanged if subsets merge or if one subset splits into two. Recall that the number of mutants in the population changes only if a replacement occurs along any link that connects residents with mutants or vice versa. Because the Moran process essentially selects links with a probability proportional to the link weight and the fitness of the node at its tail, it follows that for each subset, the probability that another mutant is added is simply given by $r/(1+r)$ and the complementary probability that one is removed is $1/(1+r)$. Since this holds for each subset, it also holds for the entire population and is independent of the number, size, shape and distribution of mutant subsets. This invariance applies if and only if the circulation theorem is satisfied. Consequentially, the fixation probability on circulation graphs reduces to the recursive Eq. (3) derived for the original Moran process.

Note that even though the fixation probabilities remain unchanged on circulation graphs, the corresponding fixation times are very sensitive to the details of the population structure and pose a much harder problem. For the original Moran process it can be shown that the fixation time of an advantageous mutant with fitness $r > 1$ in a resident population is the same as the fixation time of a disadvantageous mutant with fitness $1/r$ – however, the first scenario is much more likely (Taylor et al., 2006). The circulation theorem only ensures that the ratio of the transition probabilities $T^+/T^- = r$ remains unchanged but even on circulation graphs T^+ and T^- depend not only on the number but also on the distribution of mutants. Generally, population structures tend to substantially increase the fixation times because the structure limits the possibilities for mutants to conquer new nodes.

The circulation theorem indeed covers a large class of population structures and certainly includes the most intuitive cases such as regular lattices, cycles and fully connected graphs (see Fig. 3). Thus, it is not so surprising that Maruyama (1970) and Slatkin (1981) arrived at the conclusion that introducing population structure leaves fixation probabilities of mutants unaffected. However, the circulation theorem not only allows to determine for which population structures this applies but it also indicates that other structures must exist that *do* result in different fixation probabilities. In particular, what structures suppress selection and enhance random drift ($\rho < \rho_1$ for $r > 1$)? Do they exist and what are their characteristics? And conversely, is it possible to achieve the opposite and generate structures that enhance selection and suppresses random drift ($\rho > \rho_1$ for $r > 1$)?

2.2 Evolutionary Suppressors

Tilting the balance between selection and random drift in favor of random drift means that altering the population structure results in smaller fixation probabilities ρ of an advantageous mutant ($r > 1$) than in the original Moran process or on a circulation graph, $\rho < \rho_1$. Conversely it means that a disadvantageous mutant ($r < 1$) has a higher fixation probability, $\rho > \rho_1$. Indeed, this holds whenever a population is arranged in a hierarchical manner. Some examples of such evolutionary suppressors are shown in Fig. 5. The most extreme case is given by a linear chain where

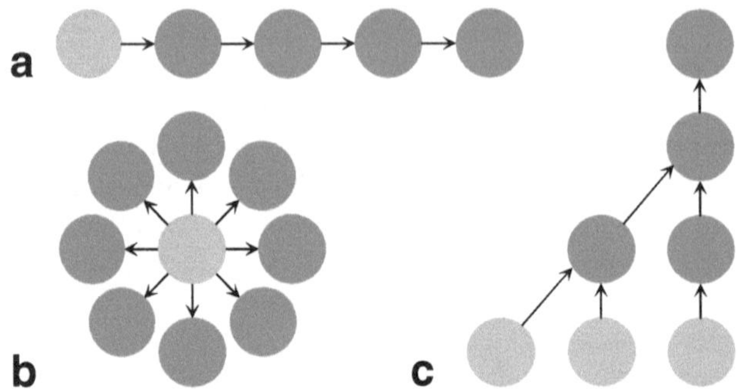

Fig. 5 Evolutionary suppressors are characterized by a hierarchical organization of the population: 'cold' nodes (light grey) are infrequently replaced (or not at all) and determine the evolutionary fate of 'hotter' downstream nodes (dark grey). **a** *Linear chain*: the offspring of every node replaces the occupant of the node to the right. The leftmost root node is never replaced and the offspring of the rightmost node is lost. **b** *Burst*: a central hub node feeds into a reservoir. The offspring of the reservoir is lost and the hub is never replaced. The linear chain and the burst structure both result in fixation probabilities of $\rho = 1/N$ – independent of the mutants fitness. A mutant fixates only if it arises in the root or hub node, respectively, but then it fixates with certainty. **c** *Multiple roots*: in this case no mutant can ever reach fixation, $\rho = 0$. If a mutation occurs in one of the root nodes, it gives rise to a persistent lineage of mutants but it cannot conquer the other root nodes

the offspring of each individual replaces the occupant of the node to its right (see Fig. 5a). The leftmost node is a root node and is never replaced whereas the offspring of the rightmost node is lost. This generates a flux through the population from left to right such that no mutant can reach fixation unless the mutation occurs in the root node. This happens with the probability $1/N$ but in that case fixation occurs with certainty. Thus the fixation is simply $\rho = 1/N$, irrespective of the mutant's fitness and hence selection is eliminated and random drift rules. If there are multiple roots (see Fig. 5c) no mutation can ever fixate, $\rho = 0$.

Evolutionary suppressors have a very simple, almost trivial structure but at the same time they turn out to be highly relevant in biological systems. While mutations enable populations to adapt to changing environments, they are generally pathogenic when they occur within an organism. Especially dangerous are those mutations that increase the net reproductive rate of a cell because this may later develop into cancer (Vogelstein and Kinzler, 1998). In order to prevent accumulation and spreading of detrimental mutations, organisms take advantage of evolutionary suppressors (Nowak et al., 2003). Epithelial tissue, such as our skin or the colon, is organized into small compartments (crypts in the colon) and each compartment is arranged in multiple layers of cells of increasing degrees of differentiation – ranging from few undifferentiated stem cells to terminally differentiated epithelial cells. With the exception of the stem cells, all cells are regularly renewed by new cells from precursor layers. This exactly matches the setup of the linear chain (see Fig. 5a) and thus cancerous mutations will be eventually washed out, unless they happen to occur in one of the stem cells. In case this occurs, then the compartmentalization confines the mutants and prevents further spreading. Another impressive example of a complex hierarchical arrangement is given by our blood system where the stem cells reside in the bone marrow and divide only about once a week through a series of precursor cells to the terminally differentiated red blood cells with a production of the order of 10^{12} cells every day. The architecture of all these systems is shaped to prevent malignant mutations from spreading.

2.3 Evolutionary Amplifiers

Evolutionary amplifiers are the counterpart to evolutionary suppressors. These population structures tilt the balance between selection and random drift in favor of selection such that the fixation probabilities ρ of advantageous mutants $(r > 1)$ is larger than in the original Moran process or on circulation graphs, $\rho > \rho_1$. Because selection is enhanced, this also implies that disadvantageous mutants $(r < 1)$ have a smaller fixation probability, $\rho < \rho_1$. Evolutionary amplifiers are also characterized by hierarchical population structures with the crucial addition of positive feedback loops. A selection of evolutionary amplifiers is shown in Fig. 6. The simplest example is given by the *star* structure where a central hub is connected to a reservoir of leaf nodes (see Fig. 6a) but in contrast to the *burst* structure (see Fig. 5b) the links between the hub and the leaves are bi-directional. The hub represents a bottleneck for the evolutionary progression because if one leaf node is occupied by a mutant

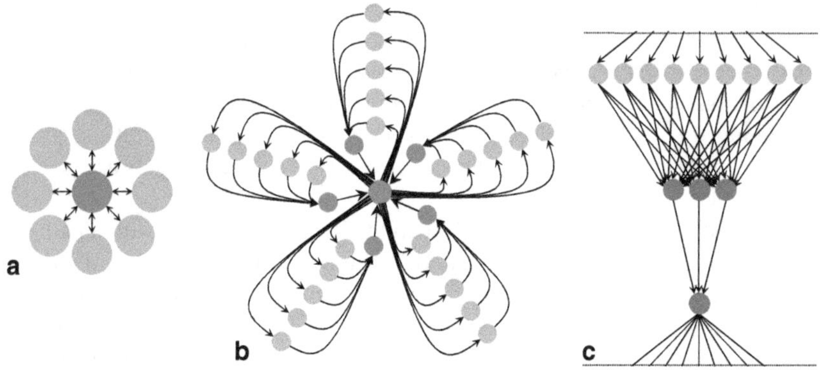

Fig. 6 Evolutionary amplifiers are characterized by a hierarchical organization of the population where 'cold' reservoir nodes (light grey) feed through a series of bottlenecks into 'hotter' (darker grey) nodes and eventually into a central hub that feeds back into the reservoir. The diameter k of the graph, i.e. the minimum number of links that connect any node with any other node, determines the amplification of the graph. A mutant with fitness r fixates with the same probability as another mutant with fitness r^k on a circulation graph. **a** *Star*: a central hub node is connected to a reservoir of leaf nodes, $k = 2$. **b** *Superstar*: several petals consist of a reservoir that is connected through a linear chain to a central hub that feeds back into all reservoirs, $k = 3$. **c** *Funnel*: a reservoir feeds into a smaller downstream sub-population that in turn feeds into a hub, which then feeds back into the reservoir, $k = 3$

it needs to conquer the hub before another leaf node can be taken over. Most of the time, the 'hot' hub is replaced by reproducing leaf nodes and only occasionally the hub itself reproduces and replaces a leaf node. For an advantageous mutant in a leaf node this means that compared to a resident leaf node it has a relative advantage of r to occupy the hub and similarly the mutant hub has again a relative reproductive advantage of r. Thus, the overall relative advantage of a mutant leaf node to proliferate and occupy another leaf node is r^2. Note that there is no other way for a mutant to spread through the population. As we will see, a mutant on the star structure with fitness r has the same fixation probability as a mutant with fitness r^2 on a circulation graph.

The principles of the star structure can be generalized to a series of sequential bottlenecks that further increase amplification such as provided by the *superstar* (Fig. 6b) or the *funnel* (Fig. 6c). The decisive quantity in all amplifiers is the minimum number of reproductive steps k that are required for a 'cold' reservoir node to populate another reservoir node. k determines the strength of the evolutionary amplifier. In an unstructured population, every node can be reached in one step and thus $k = 1$, on the star it takes two steps, $k = 2$, and on the superstar and funnel shown in Fig. 6 it takes three steps, $k = 3$. The latter two structures can be easily generalized to arbitrary k and the fixation probability becomes (Lieberman et al., 2005):

$$\rho_k = \frac{1 - \frac{1}{r^k}}{1 - \frac{1}{r^{kN}}}. \tag{5}$$

Thus, the fixation probability ρ_k of a mutant with fitness r on an evolutionary amplifier with strength k is the same as that of a mutant with fitness r^k in the original Moran process or on a circulation graph. In finite populations Eq. (5) is only an approximation and becomes exact for $N \to \infty$. Comparisons with simulation data is shown in Fig. 7a. The proof of Eq. (5) involves a recursive analysis of the sequential bottlenecks as sketched above for the star structure. Technical details are provided in Lieberman et al. (2005).

In the limit $N, k \to \infty$ the fixation probability even for a mutant with an arbitrarily small fitness advantage converges to one. Similarly, the elimination of a mutant with only an arbitrarily small fitness disadvantage happens with certainty. However, evolutionary amplification comes at a price, which is reflected in long fixation times. In fact, fixation times tend to infinity as fixation probability approaches certainty.

In contrast to evolutionary suppressors, population structures that amplify selection are rather complex and potentially less relevant in nature. However, one interesting exception occurs in the case of scale-free networks (Albert and Barabási, 2002). These networks are characterized by a power law degree distribution, where the degree or connectivity of a node indicates its number of neighbors. Therefore, few nodes are highly connected whereas most nodes entertain only few connections to other nodes. This seems to capture relevant features of social, technological and biological systems ranging from the network of scientific collaborations, to

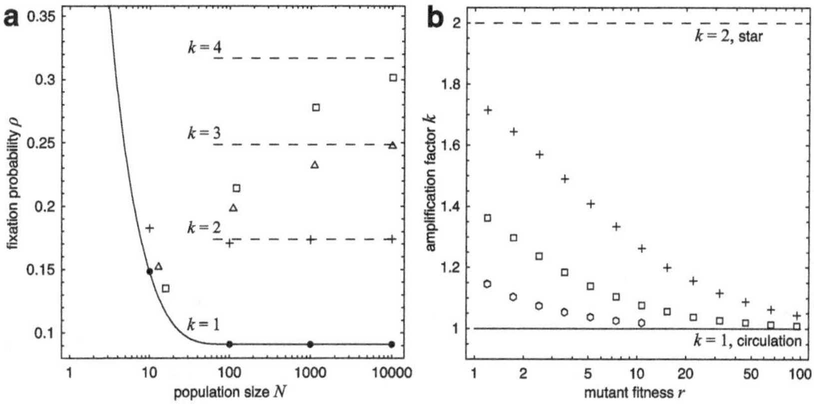

Fig. 7 Simulation data for evolutionary amplifiers. **a** fixation probabilities for an advantageous mutant with fitness $r = 1.1$ for different population structures as a function of the population size N: circulation graph (•, $k = 1$), star (+, $k = 2$) and superstars (△, $k = 3$; □, $k = 4$). The solid black line indicates the theoretical prediction for circulations and the dashed lines denote the prediction in the limit $N \to \infty$. **b** strength of amplification on scale-free networks of size $N = 100$ as a function of the mutant fitness r. The amplification decreases with increasing fitness as well as with increasing average connectivity m of the nodes: $m = 2$ (+), $m = 4$ (□) and $m = 6$ (hexagon). For marginally beneficial mutations and $m = 2$ the amplification is almost as strong as on a star structure but then gradually declines and approaches $k = 1$ for circulation graphs.

U.S. power lines (Barabasí and Albert, 1999) and gene regulatory networks (Koonin et al., 2006).

On scale-free networks the amplification factor k depends on the fitness of the mutant r (see Fig. 7b) and decreases with increasing fitness. Thus, scale-free networks selectively support mutants that are only marginally advantageous – such mutations are prone to accidental extinction through random drift – but the structure does not promote highly beneficial mutations – these fixate with very high probability anyways.

3 The Problem of Cooperation

In nature the fitness of individuals is, in general, not fixed as we have assumed so far, but instead depends on interactions with other members of the population. Evolutionary game theory (Maynard Smith, 1982; Maynard Smith and Price, 1973) provides a powerful mathematical framework to analyze situations where the performance of an individual does not only depend on its own behavior but also on the behavior of its interaction partner or opponent. The most interesting scenario refers to the evolutionary puzzle of the emergence of cooperation under Darwinian selection. The problem of cooperation is captured by social dilemmas (Dawes, 1980), which describe a conflict of interest between the community and the individual. In social dilemmas cooperators produce a valuable public good at some cost to themselves while defectors attempt to exploit the common resource without contributing themselves. Because the public good is valuable, groups of cooperators are better off than groups of defectors, but in any mixed group defectors outperform cooperators – and hence the dilemma. The most prominent game theoretical models to investigate this kind of interactions are the prisoner's dilemma and the snowdrift game for pairwise interactions as well as the public goods game for interactions in larger groups (Doebeli and Hauert, 2005). Here we focus on pairwise interactions in the prisoner's dilemma and snowdrift game and refer to Hauert et al. (2006c) for a general discussion of social dilemmas in groups of arbitrary size.

Social dilemmas are abundant in nature. For example, musk oxen create defense formations to protect their young from wolves (Hamilton, 1971). However, for each ox it would be better to avoid potential injury and to stand in the second line but if every individual behaves that way their defense breaks down and the group becomes prone to attacks by wolves. Similar conflicts of interests occur in sentinel behavior in merkats (Clutton-Brock et al., 1999), in predator inspection behavior in fish (Milinski, 1987; Pitcher, 1992), in phages competing for reproduction (Turner and Chao, 1999; Turner and Chao, 2003) or in microorganisms producing extra cellular products such as enzymes in yeast (Greig and Travisano, 2004), biofilms (Rainey and Rainey, 2003) or antibiotic resistance (Neu, 1992), to name only a few prominent examples. However, social dilemmas also occurred on an evolutionary scale and life could not have evolved without the repeated incorporation of lower level units into higher level entities. Major transitions such as the formation of chromosomes out of replicating DNA molecules, the transition from single cells to multicellular organisms or from individuals to societies require cooperation (Maynard Smith and Szathmáry, 1995). Finally, humans have taken the problem of cooperation to yet another level (Hardin, 1968) when it comes to social security such as health care or pension plans and, even more importantly, to global scales in terms of conservation natural resources such as drinking water, clean air, fisheries and our climate (Milinski et al., 2006).

3.1 Prisoner's Dilemma

The prisoner's dilemma made its first appearance in an experimental bargaining setup designed Melvin Dresher and Merill Flood in the wake of the second world war (Flood, 1958). Only later it was named by Albert W. Tucker, who contributed an illustrative anecdotal story where two burglars are arrested on the suspicion of a robbery (Poundstone, 1992). Each burglar is interrogated separately and has the options to either refuse to give evidence or to blame his fellow prisoner. They both know that if both refuse to give evidence they will be charged for a minor crime and sentenced to one year imprisonment but if they blame each other, they face three years imprisonment. However, if one refuses to give evidence but gets blamed by the other, then the first one gets the full charge of five years whereas the approver is set free. It is easy to see that no matter what the fellow prisoner decides it is always better to blame him or her in order to reduce the sentence. But if both prisoners reason the same way, they both end up three years in prison instead of only one if they had refused to give evidence. Thus, selfish interests prevent them from achieving the mutually preferable outcome. A state where none of the participants can improve its payoff by unilaterally changing its strategy is called a Nash equilibrium (Nash, 1951).

In evolutionary biology, the prisoner's dilemma is usually framed in terms of fitness costs and benefits. Cooperators provide a benefit b to their co-player at a cost c to themselves $(b > c)$ and defectors neither provide benefits nor pay costs. The payoffs for the joint behavior of two interacting individuals is usually written in the form of a payoff matrix:

$$\begin{array}{c} \\ C \\ D \end{array} \begin{array}{c} \begin{array}{cc} C & D \end{array} \\ \begin{pmatrix} b-c & -c \\ b & 0 \end{pmatrix} \end{array} \tag{6}$$

Mutual cooperation pays $b-c$ whereas mutual defection pays nothing. However if only one player defects and the other cooperates then the defectors gets the benefit b without having to pay the costs and the cooperators faces the costs c without receiving any benefit. Thus, just as in the case of the prisoners, it is always better to defect irrespective of the other players behavior but if both players follow this reasoning they end up with nothing instead of $b-c$.

Evolutionary dynamics is about populations and in this case about the change in frequencies of cooperators and defectors. In an infinite population with a fraction x cooperators $(1-x$ defectors) and randomly interacting individuals, the evolutionary fate of cooperators is given by the replicator equation:

$$\dot{x} = x(f_C - \bar{f}) = x(1-x)(f_C - f_D). \tag{7}$$

where f_C, f_D represent the average payoffs of cooperators and defectors, respectively, and $\bar{f} = xf_C + (1-x)f_D$ denotes the average population payoff. The average payoff of cooperators is simply $f_C = xb - c$ because in every interaction they pay the costs of cooperation c but only if they meet another cooperator they receive the benefit b, which happens with probability x. Similarly the average payoff of defectors is $f_D = xb$. Thus, cooperators are always worse off $(f_C < f_D)$ and irrespective of their initial frequency, they will dwindle and eventually disappear. $x^* = 0$ is the only stable equilibrium. This nicely illustrates the fact that evolutionary dynamics represents a myopic optimization process: even though fitter individuals are selected in every time step, the overall fitness of the population decreases.

In finite populations, the fitness of a player is given as $1 - w + wP$, i.e. the convex combination of a baseline fitness, which is normalized to 1 for all players, and the payoff P from the prisoner's dilemma interactions. The relative importance of the two components is specified by w. For $w \to 0$, fitness differences decrease and selection becomes weak. In order to model evolution, the Moran process is equally applicable to settings where the fitness depends on the current composition of the population, i.e. if fitness is frequency dependent (Nowak et al., 2004).

In a population of size N with i cooperators, the average fitness of cooperators and defectors is given by:

$$f_C(i) = 1 - w + \frac{w}{N-\mathbf{1}}((i-1)b - (N-1)c) \tag{8a}$$

$$f_D(i) = 1 - w + \frac{w}{N-1} ib. \tag{8b}$$

Note that for the Moran process $f_C(i), f_D(i) > 0$ must always hold in order to translate fitness into probabilities of reproduction but this is easily achieved by limiting the maximum selection strength. The replicator equation does not impose similar constraints because the fitness denotes the rate of reproduction relative to the population average.

Based on Eq. (8) the transition probabilities T^+, T^- for a change to $i+1$ or $i-1$ cooperators can be derived in analogy to Sect. 1.2. However, solving the recursive equation in order to determine the fixation probability ρ_C of a single cooperator in a population of defectors is a bit more challenging (Nowak et al., 2004; Karlin and Taylor, 1975):

$$\rho_C = 1 \Bigg/ \left(1 + \sum_{k=1}^{N-1} \prod_{i=1}^{k} \frac{f_D(i)}{f_C(i)}\right). \tag{9}$$

In the weak selection limit, $w \to 0$, Eq. (9) simplifies to

$$\rho_C = \frac{1}{N} - \frac{w}{2N}(b + (N-1)c) + o(w^2) < \frac{1}{N}. \tag{10}$$

and hence evolution never favors cooperation – cooperators are doomed just as in infinite populations.

3.2 Snowdrift Game

The anecdotal story behind the snowdrift game involves two drivers on their way home caught in a blizzard and trapped on either side of a snowdrift (Sugden, 1986). Each driver has the option to remove the snowdrift and start shoveling or to remain in the cozy warmth of the car. If both cooperate and shovel, they both receive the benefit b of getting home while sharing the labour costs c $(b > c)$ but if only one shovels both still get home but the cooperator has to do all the work. If no one shovels neither gets anywhere and they have to wait for spring to melt the snowdrift or at least for the rescue team. In contrast to the prisoner's dilemma, the best strategy now depends on the co-player's decision: if the other driver shovels it is best to be lazy but when facing a lazy bum it is better to swallow the bitter pill and to start shoveling instead of remaining stuck in the snow.

The payoff matrix of the snowdrift game is given by

$$\begin{array}{c c} & \begin{array}{cc} C & D \end{array} \\ \begin{array}{c} C \\ D \end{array} & \begin{pmatrix} b-\frac{c}{2} & b-c \\ b & 0 \end{pmatrix} \end{array} \tag{11}$$

The snowdrift game has the same characteristics as the game of Chicken or the Hawk-Dove game (Maynard Smith, 1982) but these games are usually framed in terms of competitive interactions. Note that for $2b > c > b$ the snowdrift game turns into a prisoner's dilemma. For even higher costs, $c > 2b$ mutual defection becomes the mutually preferred outcome.

In infinite populations with a fraction x cooperators the evolutionary dynamics is again determined by the replicator equation (7) with $f_C = b - c(1 - x/2)$ and $f_D = xb$. In contrast to the prisoner's dilemma, $x^* = 0$ is now unstable and an interior fixed point exists with $f_C = f_D$ for $x^* = 1 - r$ and $r = c/(2b - c)$. Thus, in the snowdrift game cooperators and defectors can co-exist at some equilibrium frequency, which is determined by the costs and benefits of the game. This originates in the fact that in the snowdrift game it is always better to adopt a strategy that differs from the co-player. As a consequence $f_C > f_D$ holds if cooperators are rare $(x \to 0)$ but $f_C < f_D$ if co-operators abound and defectors are rare $(x \to 1)$. Note that at the equilibrium, the population as a whole is worse off than if everybody would cooperate $(\bar{f} = (1-r)^2(b - c/2) < b - c/2)$ – this is the hallmark of social dilemmas and another instance of myopic optimization.

In finite populations we could proceed as before and determine the fixation probability of a single cooperator in a defector population. However, because cooperators and defectors can co-exist, the fixation probability may no longer be a relevant quantity to characterize the dynamics of this system. In fact, co-existence easily leads to exceedingly long fixation times and so situations are possible where fixation of co-operation is highly likely but requires eons to happen. Thus, with the exception of extremely small populations, equilibrium properties seem more relevant to characterize the snowdrift game in infinite as well as in finite populations.

3.3 Spatial Games

In the prisoner's dilemma cooperators are doomed in the absence of supporting mechanisms. Over the last decades several mechanisms have been proposed that are capable of establishing and maintaining cooperation among unrelated individuals. The different mechanisms essentially fall into four categories: (i) conditional behavioral rules under direct or indirect reciprocity (Trivers, 1971; Nowak and Sigmund, 1998); (ii) extensions of the strategy space by allowing for punishment or voluntary participation (Clutton-Brock and Parker, 1995; Hauert et al., 2002, 2007); (iii) feedback between ecological and evolutionary dynamics

(Hauert et al., 2006a, b); or (iv) by introducing population structures (Nowak and May, 1992; Hauert and Doebeli, 2004; Ohtsuki et al., 2006).

In this section we consider the last case and return to structured populations where individuals occupy nodes on a graph (c.f. Sect. 2) such that individuals no longer interact with all members of the population. The links of the graph define the neighborhood of all individuals and their fitness is based on interactions within this local neighborhood. As before, we are interested in how population structure affects the evolutionary dynamics and the fate of cooperators and defectors, in particular. Unfortunately, this is a hard problem and is analytically intractable in general because the fitness of each individual depends on the local configuration of its neighborhood. In fact, to fully understand the evolutionary dynamics in structured populations can be even challenging for computer simulations. Despite these bleak perspectives, there are interesting and relevant exceptions that reveal new insights into the interplay of cooperation and spatial structure.

Among the most important results ranks the finding that spatial structure supports cooperation in the prisoner's dilemma (Nowak and May, 1992). In the spatial prisoner's dilemma a cooperator provides a benefit b to all of its k neighbors at a cost kc to itself. Defectors do not provide benefits and pay no costs. Thus, a cooperator with k neighbors and i cooperators among them has an average fitness of $f_C^i = 1 - w + w(ib - kc)/k$ and a defector in the same position achieves $f_D^i = 1 - w + wib/k$ where w specifies the selection pressure on the performance in the prisoner's dilemma (c.f. Eq. (8)). Structured populations enable cooperators to thrive by forming clusters and thereby they more often interact with other cooperators and, at the same time, they reduce exploitation by defectors. However, this clustering advantage is limited and requires sufficiently small cost-to-benefit ratios c/b.

There are two fundamentally different approaches to investigate effects of space on cooperation: first we consider finite populations of size N and determine the conditions under which spatial structure promotes the evolution of cooperation such that a single cooperator in a defector population has a higher fixation probability than a neutral mutant, $\rho_C > 1/N$. In the next section a surprisingly simple and general rule is derived in the limit of weak selection, $w \to 0$, based on pair approximation (Matsuda et al., 1992; van Baalen and Rand, 1998). In Sect. 3.5 we turn to equilibrium properties in situations where cooperators and defectors co-exist for long times. In particular, it turns out that space affects cooperation rather differently in the prisoner's dilemma and the snowdrift game.

3.4 The $B > C \cdot K$-Rule

In order to derive the fixation probability of a single cooperator ρ_C, the evolutionary dynamics of cooperators and defectors on a graph can again be modeled by the Moran process where the fitness of each individual depends on interactions with all

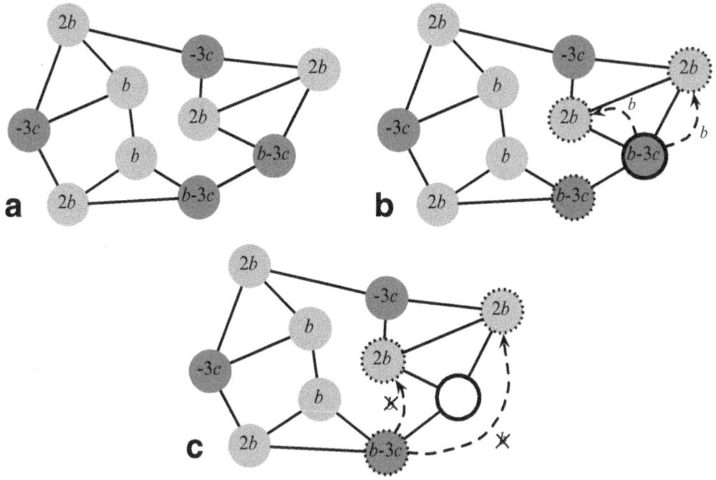

Fig. 8 Games on graphs – cooperators (dark grey) and defectors (light grey) compete on a regular graph where each node has $k = 3$ neighbors. **a** *Local interactions* – for the prisoner's dilemma, each node is marked with the fitness contribution arising from interactions with the three neighbors. **b** *Birth-death updating* – the cooperator (marked by a solid border) was selected for reproduction and its offspring will replace one of its three neighbors (dashed border). Based on their fitness, neighbors compete for reproduction but this puts cooperators at a disadvantage because they supported their defecting competitors by providing them with a benefit b (dashed arrows). **c** *Death-birth* updating – the bordered node became vacant and its neighbors (dashed border) compete to repopulate the node based on their fitness. In this case cooperators are better off because less or no support goes to the defecting competitors (dashed arrows)

other individuals in its neighborhood (Ohtsuki et al., 2006). A sample graph with prisoner's dilemma fitness values is shown in Fig. 8a. The original Moran process is formulated as a death-birth process (see Sect. 1.2): an individual is selected for reproduction with a probability proportional to its fitness and its offspring replaces a random member of the population (on graphs a random neighbor is replaced). However, the sequence of events could be easily reversed such that first a random member of the population is removed and then the remaining individuals compete to repopulate the vacant site (on graphs only the neighbors of the vacant site compete). In unstructured populations or if fitness is fixed, changing the sequence of events manifests itself in only marginal changes of the results. However, in the current setup it turns out to be of crucial importance. In fact, for the birth-death process structured populations are unable to promote cooperation and $\rho_C < 1/N$ always holds (Ohtsuki et al., 2006). The intuitive reason is that neighboring individuals compete for reproduction, which means that cooperators actually support their defecting competitors (see Fig. 8b). In contrast, for the death-birth process the disadvantage of cooperators is reduced because the individuals competing to repopulate the vacant site are typically not neighbors (even though they can be) and thus cooperators rarely feed their competitors (see Fig. 8c). Indeed, we shall see that for the death-birth process, evolution can favor cooperation if $b > c\,k$ holds.

3.4.1 Pair Approximation

Analytical approximations of the fixation probability ρ_C are based on pair approximation, which requires regular *homogeneous* graphs. On such graphs each individual has the same number of neighbors k, all links are undirected and have identical weights. Thus, the graph looks the same when viewed from any one node. This holds for any lattice (see Fig. 3 for a square lattice) as well as for less uniform structures such as random regular graphs (Bollobás, 1995) – an example with $k = 3$ is shown in Fig. 8a.

In unstructured populations the evolutionary change is described by the change in the number or frequency of cooperators (see Sect. 3.1). In structured populations this is insufficient because it neglects local correlations but it is possible to track nearest neighbor correlations using pair approximation (Matsuda et al., 1992; van Baalen and Rand, 1998). For homogeneous graphs, the pair configuration probability or the frequency of a strategy pair $p_{ss'}$ indicates the chance when randomly picking an individual and one of its neighbors that the individual has strategy s and the neighbor s'. For our purposes s, s' are either cooperation c or defection d. p_s is simply the frequency of strategy s with $p_s = p_{sc} + p_{sd}$ and because $p_c + p_d = 1$ it follows that $p_{cd} = p_{dc}$. Configuration probabilities of arbitrarily large clusters are approximated by pair configuration probabilities. For example, the probability of the three-cluster s, s', s'' is given by $p_{ss's''} = p_{ss'} \cdot p_{s's''}/p_{s'}$ where the denominator corrects for the fact that both, $p_{ss'}$ and $p_{s's''}$ include the probability for s'. Note that this approximation applies to tree graphs (or Bethe lattices) but neglects loops which are particularly important on lattices. Also note that pair approximation cannot distinguish any graphs with the same k. Since random regular graphs are locally similar to trees (Bollobás, 1995), predictions based on pair approximation are expected to be better for random regular graphs than for lattices.

The spatial dynamics can thus be approximated by four variables p_{cc}, p_{cd}, p_{dc} and p_{dd} but because they must add up to 1 and because of $p_{cd} = p_{dc}$ this only requires two dynamical equations. The most interesting quantities are the overall fraction of cooperators $p_c = p_{cc} + p_{cd}$ as well as the local affinity of cooperators, i.e. the conditional probability that the neighbor of a cooperator is another cooperator $q_{c|c} = p_{cc}/p_c$. In the following the dynamics is expressed in terms of these two quantities but to keep the formulas simple, another local quantity is sometimes used: $q_{c|d} = p_{cd}/p_d = (1 - p_c(2 - q_{c|c}))/(1 - p_c)$, i.e. the conditional probability that the neighbor of a defector is a cooperator.

3.4.2 Dynamical Equations

The change in p_c and $q_{c|c}$ is determined by the evolutionary dynamics and in this case the Moran process. Thus we need to determine the probability that a cooperator is replaced by a defector (or vice versa) as well as the effects of such a replacement on p_c and $q_{c|c}$. In the death-birth process, if a defector was eliminated, the neighborhood of the vacant site consist of k_c cooperators with probability

$$\binom{k}{k_c} q_{c|d}^{k_c}(1-q_{c|d})^{k-k_c}. \tag{12}$$

The neighboring cooperators and defectors have an average fitness of

$$f_c = 1-w+w\left((k-1)q_{c|c}\,b-k\,c\right) \tag{13a}$$

$$f_d = 1-w+w\left((k-1)q_{c|d}\,b\right). \tag{13b}$$

Note that each neighbor had at least one defector (the now vacant site) in its own neighborhood. Thus, the probability that the offspring of a cooperator succeeds in repopulating the vacant site becomes

$$\frac{k_c f_c}{k_c f_c + (k-k_c) f_d}. \tag{14}$$

If the defector is replaced by a cooperator, this increments p_c by $1/N$ and $q_{c|c}$ by $2k_c/(kN)$. The total increments are given by the sum over all $k_c = 0,\ldots,k$. Decrements in p_c and $q_{c|c}$ arise from replacing a cooperator by a defector and are given by an analogous calculation. After some algebra we obtain the leading terms in w of the dynamical equations:

$$\begin{aligned}\dot{p}_c = w\cdot\frac{k-1}{kN}p_c(1-q_{c|c})(1+q_{c|c}-q_{c|d})&\\ \times\left(b(k-1)(q_{c|c}-q_{c|d})-c\,k\right)+O(w^2)&\end{aligned} \tag{15a}$$

$$\dot{q}_{c|c} = \frac{2}{kN}(1-q_{c|c})\left[1-(k-1)(q_{c|c}-q_{c|d})\right]+O(w). \tag{15b}$$

Detailed derivations are provided in Ohtsuki et al. (2006). Equation (15) cannot be solved analytically in general but in the weak selection limit, $w \to 0$, a natural separation of time scales occurs where $q_{c|c}$ equilibrates much more quickly than p_c and thus the dynamical system rapidly converges to the slow manifold defined by $\dot{q}_{c|c} = 0$, or more explicitly by $q_{c|c} - q_{c|d} = 1/(k-1)$. This yields

$$q_{c|c} = p_c + \frac{1}{k-1}(1-p_c) \tag{16}$$

and, upon neglecting higher order terms in w, the dynamics on the slow manifold becomes

$$\dot{p}_c = w\cdot\frac{k-2}{(k-1)N}p_c(1-p_c)(b-c\,k). \tag{17}$$

Thus, in the weak selection limit, the fraction of cooperators increases provided that $b > c\,k$ holds. In order to derive the fixation probability ρ_C we assume that Eq. (16) always holds. This allows to consider a diffusion process of the random variable p_c on the slow manifold. Determining the drift and variance of the diffusion process

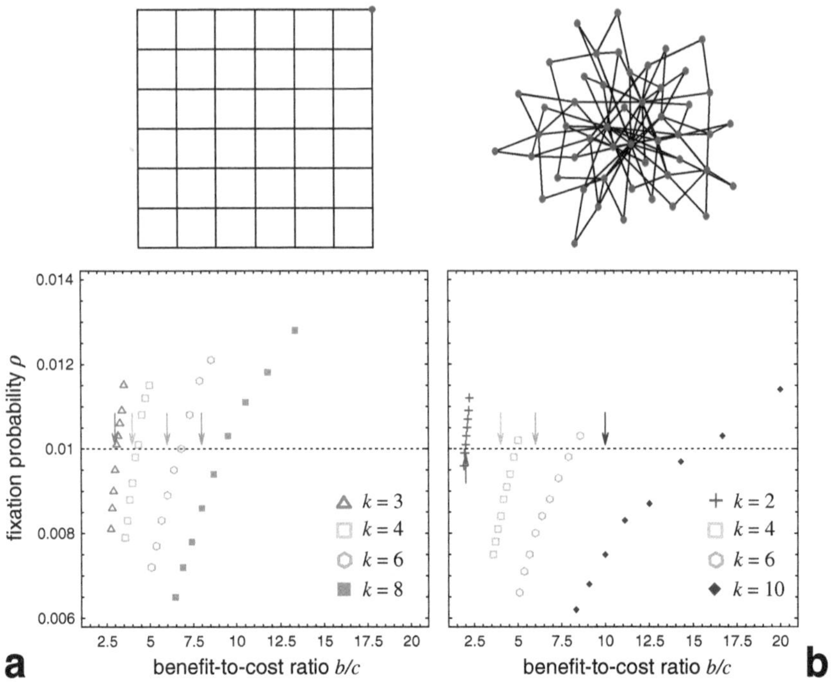

Fig. 9 Fixation probability ρ_C of a single cooperator on different types of graphs of size $N = 100$ as a function of the benefit to cost ratio b/c. The dotted horizontal line marks the fixation probability of a neutral mutant $(1/N = 0.01)$ and the arrows indicate the predictions by the $b > c\ k$-rule. **a** *Regular lattices*: deviations increase for larger k because pair approximation is based on the assumption that $N \gg k$ holds. On top, a sample lattice with $k = 4$ is shown. **b** *Scale-free networks*: good predictions are obtained even for highly inhomogeneous graphs that violate the assumptions of pair approximation. The networks are generated according to preferential attachment (Albert and Barabási, 2002) and a sample network with an average connectivity of $\langle k \rangle = 4$ is shown on the top

leads to a backward Kolmogorov differential equation for the fixation probability of a single mutant with the solution

$$\rho_c = \frac{1}{N} + w\frac{N-1}{2N}(b - k\ c). \tag{18}$$

It follows that $\rho_C > 1/N$ holds if and only if $b > c\ k$. This result is confirmed by extensive numerical simulations on various kinds of graphs (see Fig. 9). The simulations clearly show that the condition $b > c\ k$ is an excellent predictor but tends to be slightly optimistic. Moreover, $b > c\ k$ is a surprisingly robust rule and returns suitable predictions even for highly inhomogeneous graphs such as scale-free networks (Albert and Barabási, 2002) but not surprisingly the deviations tend to increase (see Fig. 9b).

In principle, the pair approximation method can be applied to any kind of game (arbitrary payoff matrices) and is not limited to the notion chosen here (Ohtsuki et al., 2006). However, if the analysis is restricted to costs and benefits of cooperation in the prisoner's dilemma, Taylor et al. (2007) have recently shown that in this special case an analysis based on inclusive fitness theory (Hamilton, 1964) naturally yields finite size corrections arising in small populations.

3.5 Equilibria

The $b > c\,k$ rule derived in the previous section determines the condition in the weak selection limit for evolution to favor cooperation in the prisoner's dilemma. In principle, an analogous calculation could be carried out for the snowdrift game but because cooperators and defectors easily co-exist in such interactions, fixation probabilities become less relevant because of exceedingly long fixation times (c.f. Sect. 3.2). The same holds in the spatial prisoner's dilemma if the selection w is sufficiently strong. Finally, in large populations the dynamics is dominated by deterministic drift and stochastic effects that are required for the fixation of one or the other strategic type becomes less relevant.

In all these cases it is more appropriate to consider large populations and investigate the effects of space on the equilibrium frequencies of cooperators and defectors in the prisoner's dilemma as well as the snowdrift game. In particular, we focus on square lattices where each individual has $k = 4$ neighbors. For the updating of the population we adopt a spatial analogue of the replicator dynamics (Hofbauer and Sigmund, 1998): a focal individual is randomly selected from the entire population and its fitness f_f corresponds to the average performance in interactions with all its neighbors. Second, a random neighbor of the focal individual is chosen and its fitness f_n is determined in the same way. The focal individual adopts the strategy of a better performing neighbor with a probability proportional to the fitness difference $z = f_n - f_f$ and sticks to its strategy otherwise. The transition probability can then be written as $\tau(z) = z^+/\alpha$, where $z^+ = z$ for $z > 0$ and zero otherwise and where α indicates a suitable normalization constant to ensure $\tau(z) \leq 1$. α depends on the type of interactions: for the prisoner's dilemma $\alpha = w(b+c)$ and for the snowdrift game $\alpha = wb$. Note that $\tau(z)$ is independent of the selection strength w and thus we set $w = 1$ without loss of generality. This particular functional form of $\tau(z)$ was chosen because it recovers the replicator Eq. (7) in the limit $N, k \to \infty$. Alternatively, all individuals could be updated in synchrony to model populations with non-overlapping generations but for stochastic update rules this barely affects the equilibrium frequencies (Doebeli and Hauert, 2005).

Unfortunately it is impossible to solve the evolutionary dynamics of this system and we have to resort to simulation data. However, pair approximation provides again a welcome analytical complement and yields useful numerical estimates. Essentially by following the reasoning in the previous section and taking the different updating procedure into account, some algebra leads to the following dynamical equations:

$$\dot{p}_c = k\, p_c^{k-1} \sum_{i=0}^{k-1} \binom{k-1}{i} q_{c|d}^i (1-q_{c|d})^{k-1-i} \sum_{j=0}^{k-1} q_{c|c}^j (1-q_{c|c})^{k-1-i}$$
$$\times \left(\tau(f_d^{i+1} - f_c^j) - \tau(f_c^j - f_d^{i+1})\right) \quad (19a)$$
$$\dot{q}_{c|c} = p_c^{k-2} \sum_{i=0}^{k-1} \binom{k-1}{i} q_{c|d}^i (1-q_{c|d})^{k-1-i} \sum_{j=0}^{k-1} q_{c|c}^j (1-q_{c|c})^{k-1-i}$$
$$\times \left((2(i+1) - k\, q_{c|c})\tau(f_d^{i+1} - f_c^j) - (2j - k\, q_{c|c})\tau(f_c^j - f_d^{i+1})\right) \quad (19b)$$

where f_c^i, f_d^i denote the fitness of cooperators and defectors that have i cooperators among their k neighbors. Also recall that $q_{c|d} = (1-p_c(2-q_{c|c}))/(1-p_c)$. Note that the sums run only up to $k-1$ because each focal cooperator must have at least one defecting neighbor and vice versa – only in these cases changes in p_c and $q_{c|c}$ can occur. Technical details on Eq. (19) are provided in Hauert and Doebeli (2004) and Hauert and Szabó (2005).

As mentioned earlier, in the prisoner's dilemma spatial structure supports cooperation (see Fig. 10). For sufficiently low cost to benefit ratios of mutual cooperation $r = c/(b-c)$ cooperators and defectors co-exist in a dynamical equilibrium. Cooperators persist by forming compact clusters such that they are more likely to interact with other cooperators and thereby reduce exploitation by defectors. However, the

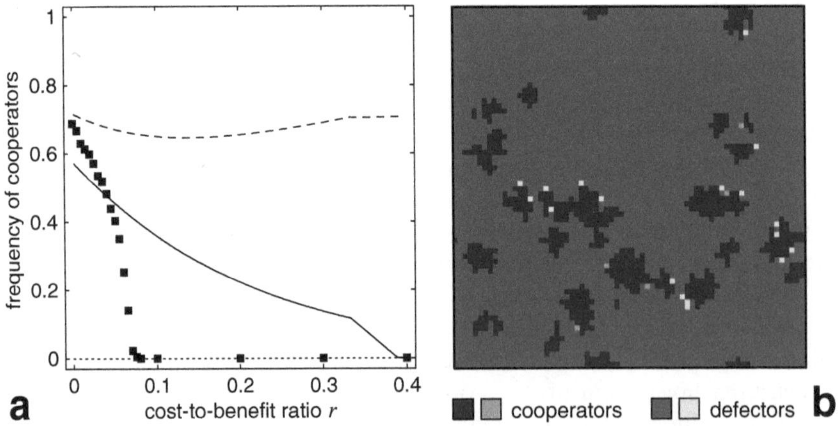

Fig. 10 Spatial prisoner's dilemma on a square 100×100 lattice with von Neumann neighborhood, $k = 4$. **a** Simulated equilibrium fraction of cooperators (solid squares) as a function of the cost-to-benefit ratio of mutual cooperation $r = c/(b-c)$. For small r cooperators persist but disappear for $r > r_c \approx 0.076$. In unstructured populations, cooperators could never survive (dotted line). Pair approximation correctly predicts the increase in cooperation due to spatial structure but it greatly overestimates its effect (p_c, solid line; $q_{c|c}$, dashed line). The consistently high values of $q_{c|c}$ indicates high degrees of clustering. **b** Snapshot of a typical lattice configuration near the extinction threshold r_c

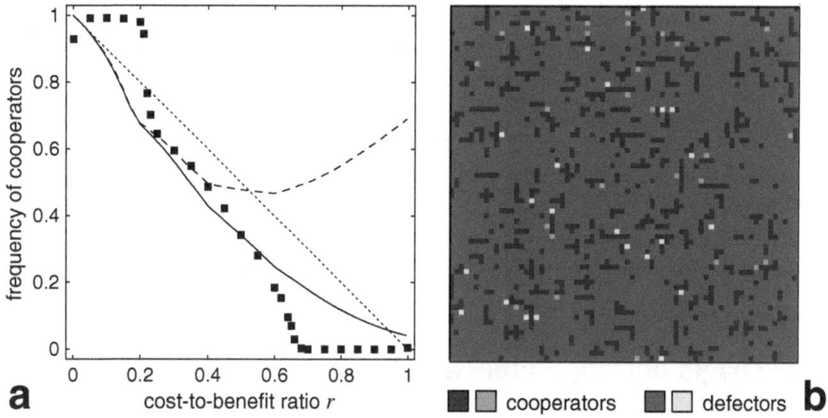

Fig. 11 Spatial snowdrift game on a square 100 × 100 lattice with von Neumann neighborhood, $k = 4$. **a** Simulated equilibrium fraction of cooperators (solid squares) as a function of the cost-to-benefit ratio of mutual cooperation $r = c/(2b - c)$. In unstructured populations cooperators and defectors co-exist (dotted line). With the exception of small r, spatial structure inhibits cooperation and for $r > r_c \approx 0.68$ space even eliminates cooperation altogether. The overall trend to inhibit cooperation is correctly predicted by pair approximation (p_c, solid line; $q_{c,c}$, dashed line) but it is unable to capture the extinction of cooperators and defectors respectively. **b** Snapshot of a typical lattice configuration near the extinction threshold r_c

clustering advantage is limited and cooperators go extinct if the benefits do not exceed the 14-fold costs. Qualitatively these results also hold for updating mechanisms based on the Moran process with marginal changes for birth-death updating and enhanced support of cooperation for death-birth updating (Hauert, 2006). Note that for the update rule inspired by the replicator dynamics $\rho_C < 1/N$ always holds, which suggests that evolution should never favor cooperation. This is no contradiction because although cooperators may never manage to reach fixation, they can nevertheless co-exist with defectors for arbitrarily long times.

Near the extinction threshold small clusters of cooperators slowly meander in a sea of defectors. Occasionally two clusters collide and merge or one cluster splits into two. This resembles a branching and annihilating random walk (see Fig. 10b) and, indeed, as r approaches the extinction threshold r_c, the simulation data suggests that the system undergoes a critical phase transition that belongs to the directed percolation universality class (Szabó and Hauert, 2002a, b).

Based on these results for the prisoner's dilemma, it was generally accepted that spatial structure promotes cooperation. However, this is not true in general for the snowdrift game. Quite on the contrary, spatial structure often inhibits the evolution of cooperation and may even eliminate cooperation altogether (see Fig. 11). In unstructured populations cooperators and defectors co-exist in a stable equilibrium because the rare type is always favored but the very same mechanism turns out to be detrimental to cooperation in spatial settings. For every individual it is always better to adopt a strategy that is different from its neighbors and this prevents the formation of larger clusters. Instead, cooperators form dendritic structures and filament-like clusters that increase interactions between cooperators and defectors (see Fig. 11b). In order to estimate the extinction threshold of cooperators we consider the threshold where the fitness of an isolated cooperator drops below the fitness of the neighboring defectors: $k(b-c) < b$, which translates to $r > (k-1)/(k+1)$. For $k = 4$ this yields a threshold of 0.6, which slightly underestimates the extinction threshold derived from simulations with $r_c = 0.68$ (see Fig. 11a). For updating mechanisms based on the Moran process the detrimental effects of space are weaker and almost disappear for death-birth updating (Hauert, 2006).

4 The Origin of Cooperators and Defectors

In nature cooperation may not always be an all or nothing decision as we have assumed in the previous section by considering the evolutionary fate of two distinct strategic types, the cooperators and the defectors. Instead, in many situations it may be more appropriate to assume a continuous range of degrees of cooperation, such as time and effort expended in providing benefits to specific individuals or for the common good. In such continuous games the strategy or trait x of an individual indicates the effort or investment in cooperative interactions that can vary between zero and an upper limit $x_{\max}$. The associated fitness benefits and costs are specified by two functions $B(x)$ and $C(x)$. We assume that $B(x)$, $C(x)$ are smooth and strictly increasing functions in the interval $[0, x_{\max}]$, with $B(0) = C(0) = 0$ such that zero investments into cooperation (or pure defection) incur no costs and provide no benefits.

In the traditional prisoner's dilemma, cooperators provide a benefit b to their partner at some cost c to themselves. Translating this setup to continuous strategies yields the payoff to an x-strategist interacting with a y-strategist: $Q(x,y) = B(y) - C(x)$. The benefits are determined by the opponents strategy whereas the costs depend on the individuals own strategy. This situation applies, for example, in grooming baboons (Saunders and Hausfater, 1988, Stammbach and Kummer, 1982) where one individual grooms the other for a time x and vice versa for a time y. In this case, the only way for an individual to improve its payoff is to reduce the costs and reduce the grooming time x. Consequentially, evolution selects lower investors and x readily approaches zero (Killingback and Doebeli, 2002). Cooperation disappears and defectors reign in both the traditional (c.f. Sect. 3.1) as well as the continuous prisoner's dilemma. The baboons avoid this unfortunate outcome because they ensure their partner's fidelity by taking turns in a single grooming session. However, our simple setup excludes such strategic responses and nothing prevents cooperation from disappearing. Moreover, complex behavioral patterns are only available to higher organisms and certainly do not occur in microorganisms.

In the traditional snowdrift game, cooperators also provide a benefit b to their partner but the costs c are shared among cooperators. Equivalently, we could assume that costs are fixed and benefits accumulate at a discounted rate $\gamma < 1$ (Hauert et al., 2006c), such that mutual cooperation yields $b(1+\gamma) - c$ and the other payoffs remain unchanged (c.f. Eq. (11)). Such situations seem to apply in yeast cells that secrete enzymes in order to hydrolyze sucrose (Greig and Travisano, 2004). If only one cell produces the enzyme the resulting food resource may be critical for survival, whereas the value of additional food is discounted because the cells get saturated. Thus, if food is scarce it may be better to invest more into enzyme production and prevent starvation despite the prospects of being exploited. Conversely, if food is abundant an individual may improve its fitness by lowering enzyme production and increase reproduction. Thus, the payoff to an x-strategist interacting with a y-strategist becomes $P(x,y) = B(x+y) - C(x)$. As before, the costs are determined by the individuals own strategy but the benefits depend on the strategies of both players. In this case one could expect that strategies would evolve away from zero to an intermediate level provided that $B(x) > C(x)$ holds for small x. However, as we shall see, the continuous snowdrift game exhibits much richer evolutionary dynamics (Doebeli et al., 2004).

The continuous snowdrift game potentially applies whenever individuals produce a valuable common resource at some cost to themselves (Doebeli and Hauert, 2005), which describes a particular but abundant form of social dilemmas. Numerous examples can be found in the microbial world ranging from viruses where replication enzymes represent a common resorce (Huang and Baltimore, 1977), and RNA phages producing proteins for the virus capsule (Turner and Chao, 2003), to antibiotic resistance in bacteria secreting β-lactamase to inhibit cell wall synthesis (Neu, 1992) and the formation of fruiting bodies in *Myxococcus xanthus* (Velicer et al., 2000). Examples from higher organisms include sentinel behavior in meerkats (Clutton-Brock et al., 1999) and predator inspection in fish (Milinski, 1987; Pitcher, 1992), where the information that the inspectors obtain can be viewed as a public resource (Magurran and Higham, 1988), to cultural evolution in humans with interactions from communal enterprises to global sustainability issues (Milinski et al., 2006).

4.1 Adaptive Dynamics

In the continuous snowdrift game the evolution of the trait x can be analyzed using the framework of adaptive dynamics (Dieckmann and Law, 1996; Geritz et al., 1998; Metz et al., 1996). Assume a homogeneous monomorphic population of x-strategists and determine whether a rare mutant with strategy y can invade. The fitness of the y strategy is simply given by $P(y,x)$ because at least as long as y-strategists are rare, interactions with other y-strategists, $P(y,y)$, can be neglected. From replicator dynamics it follows that y-strategists increase in abundance if their fitness exceeds the fitness of the resident, $P(x,x)$. Thus, the growth rate of the y-strategist is given by $f_x(y) = P(y,x) - P(x,x) = B(x+y) - C(y) - (B(2x) - C(x))$ and is called the invasion fitness because if $f_x(y) > 0$ the y mutant invades and disappears if $f_x(y) < 0$. In the limit of small mutations where y is very similar to x, it can be proven that if $f_x(y) > 0$ holds, the y-strategist not only invades but also replaces the resident population (Geritz et al., 1998). The occasional appearance of a rare mutant drives the evolutionary process but mutation rates must be small such that between subsequent invasion attempts the population has sufficient time to relax into a homogeneous state. Under these conditions, the evolution of the trait x is governed by the selection gradient $D(x) = \partial f_x/\partial y|_{y=x} = B'(2x) - C'(x)$ and the adaptive dynamics of x is described by $\dot{x} = mD(x)$ where m depends on the population size and reflects the mutational process supplying new mutant strategies. For constant population sizes m is simply a constant and is set to $m = 1$ without loss of generality. If $D(x) > 0$ mutants with $y > x$ can invade and the resident trait increases over time. Conversely, if $D(x) < 0$ mutants with $y < x$ invade and the resident trait decreases.

Equilibrium points of the adaptive dynamics, $\dot{x} = 0$, are called singular strategies x^* and are solutions to $D(x^*) = B'(2x^*) - C'(x^*) = 0$. If no such solution exists in the interval $(0, x_{\max})$, then trait values either decrease until cooperative contributions vanish $(D(x) < 0)$ or keep increasing until $x_{\max}$ is reached $(D(x) > 0)$. Both situations can occur in the continuous snowdrift game (see Fig. 12): the first case is dynamically equivalent to the continuous prisoner's dilemma whereas in the second

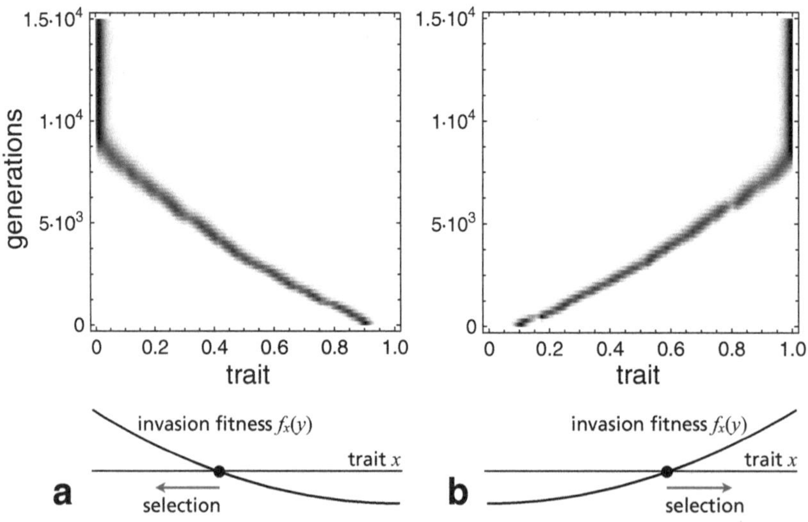

Fig. 12 Dynamics in the continuous snowdrift game in the absence of singular strategies. The top row depicts simulation results for the trait distribution in the population over time, where darker shades indicate higher abundance of traits, and the bottom row provides a schematic illustration of the fitness profile in the population. The trait interval is restricted to [0, 1] and the benefit and cost functions are quadratic, $B(x) = b_2x^2 + b_1x$, $C(x) = c_2x^2 + c_1x$, such that $C(x), B(x)$ are saturating and strictly increasing in [0, 1]. **a** The selection gradient is always negative, $D(x) < 0$, and irrespective of the initial configuration evolution keeps favoring individuals that invest less into cooperation until cooperation vanishes, just as in the continuous prisoner's dilemma. The qualitative features of the invasion fitness $f_x(y)$ do not change as x changes over time. **b** This is the exact opposite of **a**: $D(x) > 0$ always holds and the traits in the population invariably approach the maximum level of cooperation. Parameters: $b_2 = -1.5, b_1 = 7, c_2 = -1$ and **a** $c_1 = 8$; **b** $c_1 = 2$

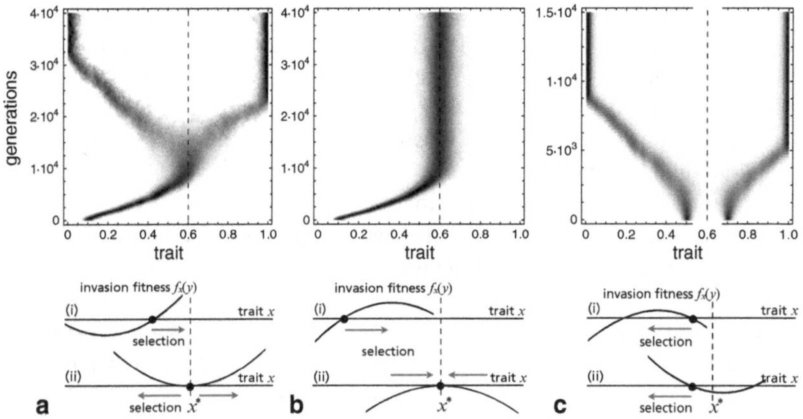

Fig. 13 Dynamics in the continuous snowdrift game in presence of a unique singular strategy x^*. The top row depicts simulation results for the trait distribution in the population over time, where darker shades indicate higher abundance of traits, and the bottom row provides a schematic illustration of the fitness profile in the population. x^* is marked by a vertical dashed line. As in Fig. 12, the trait interval is [0, 1] and the benefit and cost functions are quadratic, saturating and strictly increasing in [0, 1]. **a** *Evolutionary branching* – the singular strategy is convergent stable and the trait distribution approaches x^* but it is not evolutionarily stable and the population branches into two distinct phenotypic clusters. Evolution selects individuals with higher fitness (bottom panel (i)) but this also changes the profile of the invasion fitness $f_x(y)$ such that the fitness minimum catches up at x^* (bottom panel (ii)) and mutants with both higher and lower y can invade. **b** *Evolutionary stability* – the singular strategy is not only convergent stable but also evolutionarily stable such that the trait distribution approaches x^* and remains there. As the population converges to x^* (bottom panel (i)) the profile of $f_x(y)$ changes and at x^* the trait catches up with the maximum of $f_x(y)$ (bottom panel (ii)) and no mutants are able to invade. **c** *Evolutionary repellor* – the singular strategy is an evolutionary repellor such that the traits evolve away from x^*. Two separate simulation runs are shown: when starting below x^* cooperation disappears but if initial cooperative contributions are sufficiently high they keep increasing until the maximum is reached. In this case it is irrelevant whether x^* is evolutionarily stable (bottom panel (i)) or an evolutionary branching point (bottom panel (ii)) because evolution never reaches x^* and would require careful preparations of the initial configuration. Parameters: **a** $b_2 = -1.4, b_1 = 6, c_2 = -1.6, c_1 = 4.56$; **b** $b_2 = -1.5, b_1 = 7, c_2 = -1, c_1 = 4.6$; **c** $b_2 = -0.5, b_1 = 3.4, c_2 = -1.5, c_1 = 4$

case full cooperation is established and is sometimes termed by-product mutualism because increasing cooperation yields a net benefit to the actor and benefits to others occur only as a side effect (Connor, 1996; Dugatkin, 1996; Milinski, 1996). The dynamics becomes more interesting if x^* exists. The singular strategy is convergent stable and hence an attractor of the evolutionary dynamics if $dD(x)/dx|_{x=x^*} < 0$ and is an evolutionary repellor if this inequality is reversed, i.e. the trait x evolves away from x^* (see Fig. 13c). If x^* is an attractor, the traits in the population converge to x^* but the subsequent evolutionary fate of the population depends on whether x^* is also evolutionarily stable, i.e. whether x^* denotes a maximum or minimum of the invasion fitness $f_x(y)$. If $\partial^2 f_x(y)/\partial y^2|_{y=x^*} = 2B''(2x^*) - C''(x^*) < 0$ then x^* represents a fitness maximum and thus represents an evolutionary end state where every individual provides equal intermediate cooperative contributions and corresponds to the original expectation (see Fig. 13b). If, however, $2B''(2x^*) - C''(x^*) > 0$ then a population of x^*-strategists is at a fitness minimum and mutants with either higher or lower traits y can invade. In this case the population undergoes evolutionary branching and spontaneously splits into two distinct phenotypic clusters of high and low investing individuals (see Fig. 13a).

For quadratic benefit and cost functions $B(x) = b_2x^2 + b_1x$, $C(x) = c_2x^2 + c_1x$ with suitable parameters, such that both benefits and costs are strictly increasing over the trait interval $[0, x_{\max}]$, all dynamical scenarios occur (see Figs. 12, 13) and can be fully analyzed (Doebeli et al., 2004). In this case, the singular strategy x^* is unique (if it exists) and is given by $x^* = (c_1 - b_1)/(4b_2 - 2c_2)$. It is convergent stable if $2b_2 - c_2 < 0$ and evolutionarily stable if $b_2 - c_2 < 0$. The existence of x^* requires either (i) $4b_2 - 2c_2 > c_1 - b_1 > 0$ or (ii) $4b_2 - 2c_2 < c_1 - b_1 < 0$. In the first case x^* is always a repellor and whether it is evolutionarily stable is irrelevant because the singular strategy is never reached from generic initial conditions. In the second case x^* is always convergent stable and if, in addition, $b_2 - c_2 < 0$ holds it is also evolutionarily stable. Finally, if $2b_2 < c_2 < b_2 < 0$ then x^* is an evolutionary branching point. In order to analyze the dynamics after branching has occurred, the invasion fitness needs to be derived for a third strategic type z attempting to invade a resident population where the two co-existing strategies $x > x^* > y$ are in equilibrium. This leads to two dynamical equations one for each branch x and y (Doebeli et al., 2004). In the case of quadratic cost and benefit functions the traits keep diverging until they reach the boundaries of the trait interval because mutants with either $z > x$ or $z < y$ can invade.

Note that the phenotypic diversification occurs spontaneously and in populations with randomly interacting individuals and thus does not rely on any kind of assortment in terms of mating preferences or spatial segregation. The evolutionary end state consist of a population with pure defectors and pure cooperators that, in fact, engage in traditional snowdrift game interactions (c.f. Sect. 3.2). Thus, the continuous snowdrift game suggest an evolutionary pathway for social diversification and for the origin of cooperators and defectors.

4.2 Two Tragedies

The conflict of interest in social dilemmas is equivalently captured by the *Tragedy of the Commons* (Hardin, 1968), which states that public resources are bound to be overexploited. Especially in the context of humans, this comes at no surprise – Aristotle (384–322 BC) already drew the same conclusion: "That which is common to the greatest number has the least care bestowed upon it." The spontaneous diversification into co-existing high and low investors in the continuous snowdrift game may equally apply in communal enterprises in humans and generate, in addition, the *Tragedy of the Commune* (Doebeli et al., 2004), which states that evolution may not favor egalitarian contributions to the common good but instead promote highly asymmetric involvements. But, large differences in cooperative contributions bear a formidable potential for escalating conflicts based on the accepted notion of fairness.

5 Conclusions

This brief review discusses different approaches to model evolutionary dynamics and address the problem of cooperation. Despite the simple principles underlying evolutionary models, they often exhibit rich, complex and sometimes even chaotic but always truly fascinating dynamics. Because of this, it can be quite challenging to develop an intuitive understanding of the dynamical features. In order to achieve this intuition it is very helpful to study characteristics of spatio-temporal patterns or to follow the evolutionary change of strategy distributions in a population. This is facilitated by the *VirtualLabs* (Hauert, 2007), which complement the research results presented in this review. The growing collection of interactive on-line tutorials comprises at least one tutorial for each section covered in this review. Based on Java applets, most of the results reported here can be easily reproduced and further explorations are encouraged by the possibility to change various settings. But it is also possible to simply watch and enjoy the hypnotizing beauty of evolutionary kaleidoscopes – find out what they are by visiting the *VirtualLabs*.

Acknowledgements Support by the John Templeton Foundation is gratefully acknowledged. The Program for Evolutionary Dynamics (PED) at Harvard University is sponsored by Jeffrey Epstein.

References

Albert, R. and Barabási, A.-L. (2002) Statistical mechanics of complex networks, *Rev. Mod. Phys.* **74**, 47–97.

Barabasí, A.-L. and Albert, R. (1999) Emergence of scaling in random networks, *Science* **286**, 509–512.

Bollobás, B. (1995) *Random Graphs*, New York, Academic.

Clutton-Brock, T. H., M. J. O'Riain, P. N. M. B., Gaynor, D., Kansky, R., Griffin, A. S., and Manser, M. (1999) Selfish sentinels in cooperative mammals, *Science* **284**, 1640–1644.

Clutton-Brock, T. H. and Parker, G. A. (1995) Punishment in animal societies, *Nature* **373**, 209–216.

Connor, R. C. (1996) Partner preferences in by-product mutualisms and the case of predator inspection in fish, *Anim. Behav.* **51**, 451–454.

Dawes, R. M. (1980) Social dilemmas, *Ann. Rev. Psychol.* **31**, 169–193.

Dieckmann, U. and Law, R. (1996) The dynamical theory of coevolution: a derivation from stochastic ecological processes, *J. Math. Biol.* **34**, 579–612.

Doebeli, M. and Hauert, C. (2005) Models of cooperation based on the Prisoner's Dilemma and the Snowdrift game, *Ecol. Lett.* **8**, 748–766.

Doebeli, M., Hauert, C., and Killingback, T. (2004) The evolutionary origin of cooperators and defectors, *Science* **306**, 859–862.

Dugatkin, L. A. (1996) Tit for tat, by-product mutualism and predator inspection: a reply to Connor, *Anim. Behav.* **51**, 455–457.

Flood, M. (1958) Some experimental games, *Management Science* **5**, 5–26.

Geritz, S. A. H., Kisdi, E., Meszéna, G., and Metz, J. A. J. (1998) Evolutionarily singular strategies and the adaptive growth and branching of the evolutionary tree, *Evol. Ecol.* **12**, 35–57.

Greig, D. and Travisano, M. (2004) The Prisoner's Dilemma and polymorphism in yeast SUC genes, *Biol. Lett.* **271**, S25–S26.

Hamilton, W. D. (1964) The genetical evolution of social behaviour I, *J. Theor. Biol.* **7**, 1–16.

Hamilton, W. D. (1971) The geometry of the selfish herd, *J. Theor. Biol.* **31**, 295–311.

Hardin, G. (1968) The tragedy of the commons, *Science* **162**, 1243–1248.

Hauert, C. (2006) Spatial effects in social dilemmas, *J. Theor. Biol.* **240**, 627–636.

Hauert, C. (2007) VirtualLabs: Interactive Tutorials on Evolutionary Game Theory, http://www.univie.ac.at/virtuallabs.

Hauert, C., De Monte, S., Hofbauer, J., and Sigmund, K. (2002) Volunteering as red queen mechanism for cooperation in public goods games, *Science* **296**, 1129–1132.

Hauert, C. and Doebeli, M. (2004) Spatial structure often inhibits the evolution of cooperation in the Snowdrift game, *Nature* **428**, 643–646.

Hauert, C., Holmes, M., and Doebeli, M. (2006a) Evolutionary games and population dynamics: maintenance of cooperation in public goods games, *Proc. R. Soc. Lond. B* **273**, 2565–2570.

Hauert, C., Holmes, M., and Doebeli, M. (2006b) Evolutionary games and population dynamics: maintenance of cooperation in public goods games, *Proc. R. Soc. Lond. B* **273**, 3131–3132.

Hauert, C., Michor, F., Nowak, M., and Doebeli, M. (2006c) Synergy and discounting of cooperation in social dilemmas, *J. Theor. Biol.* **239**, 195–202.

Hauert, C. and Szabó, G. (2005) Game theory and physics, *Am. J. Phys.* **73**, 405–414.

Hauert, C., Traulsen, A., Brandt, H., Nowak, M. A., and Sigmund, K. (2007) Via freedom to coercion: the emergence of costly punishment, *Science* **316**, 1905–1907.

Hofbauer, J. and Sigmund, K. (1998) *Evolutionary Games and Population Dynamics*, Cambridge, Cambridge University Press.

Huang, A. S. and Baltimore, D. (1977) *Comprehensive Virology*, Vol. 10, Chapt. Defective Interfering Animal Viruses, pp. 73–116, New York, Plenum.

Karlin, S. and Taylor, H. M. (1975) *First Course in Stochastic Processes*, 2nd edition London, Academic.

Killingback, T. and Doebeli, M. (2002) The continuous prisoner's dilemma and the evolution of cooperation through reciprocal altruism with variable investment, *Am. Nat.* **160**, 421–438.

Koonin, E. V., Wolf, Y. I., and Karev, G. P. (2006) *Power Laws, Scale-Free Networks and Genome Biology*, Molecular Biology Intelligence Unit, Springer, New York, NY.

Lieberman, E., Hauert, C., and Nowak, M. A. (2005) Evolutionary dynamics on graphs, *Nature* **455**, 312–316.

Magurran, A. E. and Higham, A. (1988) Information transfer across fish shoals under predator threat, *Ethology* **78**, 153–158.

Maruyama, T. (1970) Effective number of alleles in a subdivided population, *Theor. Pop. Biol.* **1**, 273–306.

Matsuda, H., Ogita, N., Sasaki, A., and Sato, K. (1992) Statistical mechanics of populations, *Prog. Theor. Phys.* **88**, 1035–1049.

Maynard Smith, J. (1982) *Evolution and the Theory of Games*, Cambridge, Cambridge University Press.

Maynard Smith, J. and Price, G. (1973) The logic of animal conflict, *Nature* **246**, 15–18.

Maynard Smith, J. and Szathmáry, E. (1995) *The Major Transitions in Evolution*, Oxford, W. H. Freeman.

Metz, J. A. J., Geritz, S. A. H., Meszena, G., Jacobs, F. J. A., and van Heerwaarden, J. S. (1996) Adaptive dynamics: a geometrical study of the consequences of nearly faithful replication, In S. J. van Strien and S. M. Verduyn Lunel (eds.), *Stochastic and Spatial Structures of Dynamical Systems*, Amsterdam, North Holland, pp. 183–231.

Milinski, M. (1987) Tit for tat in sticklebacks and the evolution of cooperation, *Nature* **325**, 433–435.

Milinski, M. (1996) By-product mutualism, tit-for-tat and cooperative predator inspection: a reply to Connor, *Anim. Behav.* **51**, 458–461.

Milinski, M., Semmann, D., Krambeck, H.-J., and Marotzke, M. (2006) Stabilizing the Earths climate is not a losing game: supporting evidence from public goods experiments, *Proc. Natl. Acad. Sci. USA* **103**, 3994–3998.

Moran, P. A. P. (1962) *The Statistical Processes of Evolutionary Theory*, Oxford, UK, Clarendon.

Nash, J. (1951) Non-cooperative games, *Annals of Mathematics* **54**, 286–299.

Neu, H. C. (1992) The Crisis in Antibiotic Resistance, *Science* **257**, 1064–1073.

Nowak, M. A. (2006) *Evolutionary Dynamics*, Cambridge MA, Harvard University Press.

Nowak, M. A. and May, R. M. (1992) Evolutionary Games and Spatial Chaos, *Nature* **359**, 826–829.

Nowak, M. A., Michor, F., and Iwasa, Y. (2003) The linear process of somatic evolution, *Proc. Natl. Acad. Sci. USA* **100**, 14966–14969.

Nowak, M. A., Sasaki, A., Taylor, C., and Fudenberg, D. (2004) Emergence of cooperation and evolutionary stability in finite populations, *Nature* **428**, 646–650.

Nowak, M. A. and Sigmund, K. (1998) Evolution of indirect reciprocity by image scoring, *Nature* **393**, 573–577.

Ohtsuki, H., Hauert, C., Lieberman, E., and Nowak, M. A. (2006) A simple rule for the evolution of cooperation on graphs and social networks, *Nature* **441**, 502–505.

Pitcher, T. (1992) Who dares, wins: the function and evolution of predator inspection behavior in shoaling fish, *Neth. J. Zool.* **42**, 371–391.

Poundstone, W. (1992) *Prisoner's Dilemma*, New York, Doubleday.

Rainey, P. B. and Rainey, K. (2003) Evolution of cooperation and conflict in experimental bacterial populations, *Nature* **425**, 72–74.

Saunders, C. D. and Hausfater, G. (1988) The functional significance of baboon grooming behavior, *Annals N. Y. Acad. Sci.* **525**, 430–432.

Slatkin, M. (1981) Fixation probabilities and fixation times in a subdivided population, *Evolution* **35**, 477–488.

Stammbach, E. and Kummer, H. (1982) Individual contributions to a dyadic interaction: an analysis of baboon grooming, *Anim. Behav.* **30**, 964–971.

Sugden, R. (1986) *The Economics of Rights, Co-operation and Welfare*, Oxford and New York, Blackwell.

Szabó, G. and Hauert, C. (2002a) Evolutionary prisoner's dilemma with optional participation, *Phys. Rev. E* **66**, 062903.

Szabó, G. and Hauert, C. (2002b) Phase transitions and volunteering in spatial public goods games, *Phys. Rev. Let.* **89**, 118101.

Taylor, C., Iwasa, Y., and Nowak, M. A. (2006) A symmetry of fixation times in evolutionary dynamics, *J. Theor. Biol.* **243**, 245–251.

Taylor, P. D., Day, T., and Wild, G. (2007) Evolution of cooperation in a finite homogeneous graph, *Nature*. **447**, 469–472.

Traulsen, A., Claussen, J. C., and Hauert, C. (2005) Coevolutionary dynamics: from finite to infinite populations, *Phys. Rev. Lett.* **95**, 238701.

Trivers, R. L. (1971) The evolution of reciprocal altruism, *Q. Rev. Biol.* **46**, 35–57.

Turner, P. E. and Chao, L. (1999) Prisoner's dilemma in an RNA virus, *Nature* **398**, 441–443.

Turner, P. E. and Chao, L. (2003) Escape from prisoner's dilemma in RNA phage Φ6, *Am. Nat.* **161**, 497–505.

van Baalen, M. and Rand, D. A. (1998) The unit of selection in viscous populations and the evolution of altruism, *J. Theor. Biol.* **193**, 631–648.
Velicer, G. J., Kroos, L., and Lenski, R. E. (2000) Developmental cheating in the social bacterium Myxococcus xanthus, *Nature* **404**, 598–601.
Vogelstein, B. and Kinzler, K. W. (1998) *The Genetic Basis of Human Cancer*, Toronto, McGraw-Hill.
Watts, D. J. and Strogatz, S. H. (1998) Collective dynamics of 'small world' networks, *Nature* **393**, 440–442.

Small RNA Control of Cell-to-Cell Communication in *Vibrio Harveyi* and *Vibrio Cholerae*

Small RNA Control of Quorum Sensing

Sine Lo Svenningsen

Abstract Quorum sensing is a process of cell-to-cell communication, by which bacteria coordinate gene expression and behavior on a population-wide scale. Quorum sensing is accomplished through production, secretion, and subsequent detection of chemical signaling molecules termed autoinducers. The human pathogen *Vibrio cholerae* and the marine bioluminescent bacterium *Vibrio harveyi* incorporate information from multiple autoinducers, and also environmental signals and metabolic cues into their quorum-sensing pathways. At the core of these pathways lie several homologous small regulatory RNA molecules, the Quorum Regulatory RNAs. Small noncoding RNAs have emerged throughout the bacterial and eukaryotic kingdoms as key regulators of behavioral and developmental processes. Here, I review our present understanding of the role of the Qrr small RNAs in integrating quorum-sensing signals and in regulating the individual cells response to this information.

Keywords Quorum sensing, autoinducer, small RNAs, *Vibrio cholerae, Vibrio harveyi*

1 Introduction

Bacteria live in an environment that is constantly changing, and therefore they must constantly detect and adapt to alterations in external parameters such as nutrient availability, nutrient composition, acidity, and osmolarity, to survive. Consequently, the bacterial cell membrane abounds with sensor proteins that detect these changes in the surroundings, and cells are well equipped to interpret this information to

Sine Lo Svenningsen
Laboratory of Prof. Bonnie Bassler
Dept. of Molecular Biology
Princeton University, Princeton, NJ 08544, USA.
Email: ssvennin@princeton.edu

Arne T. Skjeltorp, Alexander V. Belushkin (eds.), Evolution from Cellular to Social Scales.

guide changes in gene expression, enabling them to adapt to the new conditions. Despite this sophisticated ability to interact with their environment, bacterial cells have traditionally been viewed as primitive, anti-social organisms, each cell living in ignorance of its neighbors. By contrast, communication between individual cells was long understood to be a prerequisite for the successful development of higher organisms. In the past fifteen years, however, it has become clear that many bacterial species engage in cell-to-cell communication, and that changes in the density and composition of the surrounding microbial community are among the key parameters sensed by an individual bacterium.

The term quorum sensing was coined to describe bacteria sensing population density and initiating new or terminating former behaviors when the cell density reached a certain threshold, a quorum.[1] In its simplest form, quorum sensing allows bacteria to exist in one of two behavioral modes: When the cell density is low they behave as individual cells, but when the cell density is high they switch into a social mode and coordinate their gene expression, allowing them to undertake tasks in unison.

Information about the population density can be valuable for making a large number of behavioral changes. *Vibrio fischeri*, for example, generates bioluminescence, which is used by its squid host for counterillumination to mask its shadow and thereby avoid predation (Visick et al., 2000). *V. fischeri* alternates between living in the high-cell-density environment of the squid light organ and the low-cell-density environment of the open ocean. By submitting the bioluminescence genes to quorum-sensing control, *V. fischeri* populations can synchronize initiation of bioluminescence once inside the host, and avoid the cost of bioluminescing when not host-associated (reviewed in Whitehead et al., 2001).

It is not only obscure symbionts in the ocean that take advantage of quorum sensing. In fact, it is now understood that quorum sensing is more the norm than the exception in the microbial world.

The hallmark of quorum sensing is the production, secretion, exchange and detection of chemical signaling molecules. As a population of quorum-sensing bacteria grows, a proportional increase in the extracellular concentration of the signaling molecule occurs. When a threshold concentration is reached, the group detects the signal and responds to it with a population-wide alteration in gene expression (Bassler and Losick, 2006).

2 Signals for Intra- and Interspecies Communication

2.1 Intraspecies Communication by AHLs and Oligopeptides

The signaling molecule of *V. fischeri* is an acyl homoserine lactone (AHL), 3OC6-homoserine lactone (Fig. 1A, Eberhard et al., 1981). Many gram-negative bacteria communicate with AHL molecules, produced by LuxI-type enzymes. The AHL's diffuse across the cell membrane, and when they reach a certain threshold

[1] Quorum: The number of members required to be present to conduct official business.

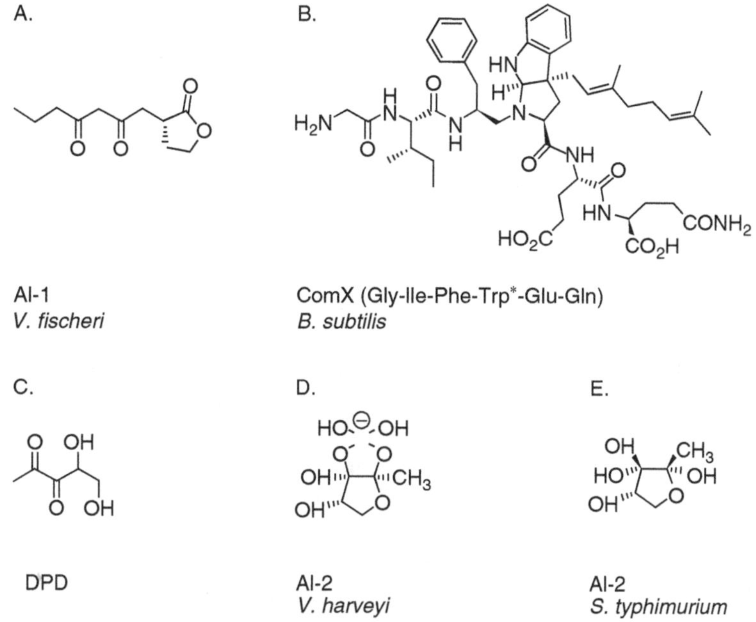

Fig. 1 Representative quorum-sensing signaling molecules. A) V. harveyi Al-1. B) B. subtilis ComX. The * above the tryptophan indicates an isoprenyl modification. C) DPD, the precursor to Al-2. D) V. harveyi Al-2. E) S. typhimurium Al-2. (Adapted from Waters and Bassler, 2005 and Bassler and Losick, 2006)

concentration, they bind a cytoplasmic transcriptional regulator of the LuxR-type. The LuxR-AHL complex binds DNA and activates/represses target genes. AHL's have various chain lengths and carry different side-chain modifications, which make the AHL produced by each species uniquely identified. The binding pocket of each species' LuxR protein is sensitive to the differences in AHL structure, so a particular species' LuxR will only recognize the AHL produced by its own kind and be unaffected by AHL produced by other species. Thus, AHL signaling is primarily used for intraspecies communication.

Gram-positive bacteria usually employ short modified oligopeptides as quorum-sensing signals. An example is ComX, a 6–10 amino acid (the length varies among *Bacillus* strains) modified oligopeptide that induces natural genetic competence in *Bacillus subtilis* at high cell density (Fig. 1B, reviewed in Tortosa and Dubnau, 1999). The peptides are often detected by membrane-bound two-component sensors that transmit the extracellular information to transcriptional regulators via a phosphorelay cascade. Like the AHL-LuxR pairs, the peptide-sensor pairs of gram-positive bacteria exhibit exquisite species-specificity.

2.2 AI-2-Mediated Interspecies Communication

AHLs and peptides represent the two major classes of quorum-sensing molecules in gram-negative and gram-positive bacteria, respectively. However, AHLs and peptides are only two of a large variety of signals employed by microorganisms. An alternative signaling molecule, called AI-2, is produced by a wide variety of bacterial species and is used to regulate many different behaviors. The marine bacterium *Vibrio harveyi*, for example, activates bioluminescence in response to AI-2 (Bassler et al., 1994), while *Clostridium perfringens* uses AI-2 to regulate toxin production (Ohtani et al., 2002) and in *Heliobacter pylori*, motility is controlled by AI-2 (Osaki et al., 2006). In every case studied, the enzyme LuxS is required for AI-2 production. The *luxS* gene is present in roughly half of all sequenced bacterial genomes (Xavier and Bassler, 2003).

AI-2 is produced from the methyl donor *S*-adenosylmethionine (SAM) (Schnauder et al., 2001). SAM is used in essential processes such as DNA, RNA and protein synthesis, so AI-2 production is tightly coupled to central metabolism. Methyl donation from SAM produces the toxic intermediate *S*-adenosylhomocysteine (SAH), which is detoxified by hydrolysis to *S*-ribosylhomocysteine (SRH) and adenine by the enzyme Pfs. LuxS catalyzes the cleavage of SRH to 4,5-dihydroxy 2,3-pentanedione (DPD, Fig. 1C) and homocysteine (Schnauder et al., 2001). Homocysteine is methylated to regenerate SAM, and the only known role of DPD is to serve as the precursor for AI-2. DPD is a highly reactive molecule, and chemical analyses of active AI-2 produced *in vitro* revealed that DPD spontaneously cyclizes to a mixture of furanone rings (Schnauder et al., 2001). Crystallography of AI-2-bound receptors from two species, the marine bacterium *V. harveyi* (Chen et al., 2002) and the enteric bacterium *Salmonella typhimurium* (Miller et al., 2004), revealed that the active moiety of AI-2 differs between species. In the case of *V. harveyi* the active form of AI-2 is a furanosyl borate diester (Fig. 1D). Boron has almost no known roles in biology, but is present at high concentrations in sea water (~0.4 mM), the natural habitat of *V. harveyi*. The AI-2 variant that binds the *S. typhimurium* receptor has different stereochemistry and does not include boron (Fig. 1E). Importantly, the two AI-2 species exist in equilibrium and interconvert readily. If the boron concentration of the growth medium is increased, AI-2 shifts to the borated form and this increases induction of bioluminescence in *V. harveyi* while it reduces the AI-2 response from *S. typhimurium*. The converse is true when the medium is depleted for boron (Miller et al., 2004).

Together, the conserved LuxS-catalyzed production of the AI-2 precursor and the environmentally dictated final fates of the product allow the economical development of a chemical vocabulary that could potentially provide bacteria with information about both the numbers and identities of their neighbors, as well as about the growth characteristics of their environment. Because behaviors regulated by AI-2 are often niche specific, it is possible that AI-2 alerts bacteria that they are in a particular environmental niche. This certainly seems to be the case for *Pseudomonas aeruginosa*, which forms thick, well-organized structures known as biofilms[2] in the lungs of cystic fibrosis patients. *P. aeruginosa* lacks the *luxS* gene, so it does not produce AI-2, but it detects AI-2 produced by other species in the lung and initiates

[2] Biofilm: A complex aggregation of microorganisms marked by the excretion of a protective and adhesive matrix.

biofilm formation in response. Presumably, AI-2 secreted from other species informs *P. aeruginosa* that it has arrived in the cystic fibrosis lung (Duan et al., 2003).

Another example of interspecies communication is found in the bacteria that make up dental plaque. Dental plaque consists of biofilms made up of hundreds of species of bacteria, two of which are *Porphyromonas gingivalis* and *Streptococcus gordonii*. *S. gordonii* is a primary colonizer, and adherence of *S. gordonii* to teeth is a prerequisite for subsequent colonization by *P. gingivalis*. In the lab, *P. gingivalis* will form a biofilm on glass slides precoated with *S. gordonii*. However, if the two species are *luxS* mutants, *P. gingivalis* does not form a biofilm. Biofilm formation can be restored if the *luxS* gene is introduced into either one of the two species (McNab et al., 2003). A recent study suggests that AI-2 also induces dual-species biofilm in two commensal human oral bacteria, *Actinomyces naeslundii* and *Streptococcus oralis*, but only when AI-2 is at an optimal concentration. Above and below this concentration biofilm-formation is suppressed (Rickard et al., 2006).

2.3 Quorum Quenching

The complexity and diversity of the quorum-sensing vocabulary presents many ways to utilize this chemical source of information to control gene expression. However, bacteria relying on quorum sensing to control critical behaviors also become vulnerable to the spread of misinformation, cheating, eavesdropping (as employed by *P. aeruginosa* in the cystic fibrosis lung), and even deadly assaults.

For example, some *Bacillus* species secrete an enzyme, AiiA, which indiscriminately cleaves the AHLs used for communication by gram-negative bacteria. This way, *Bacillus sp.* can inhibit quorum sensing in their competitors, without affecting their own peptide-based quorum-sensing system (Dong et al., 2001). In another example, four groups of *Staphylococcus aureus* have developed a scheme to favor their own kin over more distantly related *S. aureus* groups. The four groups secrete similar quorum-sensing autoinducers, but the signal molecule of one group will bind and inhibit the sensors of the other three groups, while properly transmitting information via its cognate sensor (Lyon et al., 2002). In an extreme case, *P. aeruginosa* packs its quorum-sensing signals into vesicles, allowing the hydrophobic molecules to travel through the aqueous environment until they fuse with the membrane of neighboring cells, releasing their contents into the neighbor's cytoplasm. The same vesicles used for intraspecies communication when fusing with other *P. aeruginosa*, contain molecules that function as antibiotics when released into other species of bacteria, such as *Staphylococcus epidermidis* (Mashburn and Whiteley, 2005).

It is not only other prokaryotes that take advantage of their opponent's dependence on quorum sensing to control critical behaviors. The red macroalgae *Delisea pulchra* prevents bacterial colonization of its surface by inhibiting quorum-sensing controlled biofilm formation in bacteria. *D. pulchra* coats its surface with AHL-like molecules that are internalized by the bacteria, but rather than transmitting information, the molecules trigger degradation of the quorum-sensing sensor proteins. Intriguingly, serum from humans and other mammals contain enzymes, paraoxonases, which show strong AHL-degrading activity, suggesting that quorum quenching could be an innate mammalian defense mechanism (recently reviewed in Dong et al., 2007). Presumably these behaviors have evolved for a species or group to outcompete quorum-sensing bacteria or for a eukaryotic host to fend off a quorum-sensing invader.

Quorum sensing controls virulence genes in many human pathogens. Naturally, researchers understand the potential to synthesize antimicrobial drugs that interfere with quorum sensing. Because quorum-sensing inhibitors are not lethal to the bacteria, the occurrence of bacterial resistance to the drug is likely to be reduced compared to traditional antibiotics (Hentzer and Givskov, 2003). In promising host-pathogen studies, synthetic antagonists of the quorum-sensing signals of *S. aureus* and *P. aeruginosa* have been shown to reduce virulence in mice (Mayville et al., 1999; Wu et al., 2004), and transgenic potato and tobacco plants expressing AiiA show strong resistance to the plant pathogen *Erwinia carotovora*, which uses AHL signals to activate its virulence genes (Dong et al., 2001).

3 The Quorum-Sensing Response

Identification of the chemical signals used for quorum sensing and the target genes regulated by quorum sensing in different species of bacteria provides a comprehensive understanding of the extent of cell-to-cell communication and its uses in the bacterial world. This insight is critical for understanding natural mixed-species ecosystems as well as for the design of a new generation of antimicrobials. An equally attractive research topic is how the diverse information contained in autoinducers is interpreted and transformed into optimal changes in gene expression inside the quorum-sensing bacterium. Perhaps the simplest scheme is the LuxI/LuxR-type systems described above. The LuxR-AHL complex has a much higher affinity for its target promoters than does LuxR alone. Thus, LuxR-AHL directly binds DNA and mediates the required changes in gene expression by acting as a repressor or an activator at its target promoters (Whitehead et al., 2001; Minogue et al., 2002). The scenario becomes more complicated when the signal is transmitted from membrane-bound sensors to cytoplasmic regulators and in particular when information from multiple autoinducers is integrated with additional environmental and information. A variety of quorum-sensing network architectures have been described that accomplish the goal of integrating multiple inputs and transforming them into appropriate changes in gene expression. For a recent review of the molecular arrangements used by different quorum-sensing species, see Waters and Bassler (2005). In the following, I will discuss recent advances in understanding signal integration and target gene regulation in *Vibrio harveyi* and the related human pathogen *Vibrio cholerae*.

3.1 Small RNAs Are Central to Quorum Sensing in Vibrio harveyi *and* Vibrio cholerae

Communication via multiple autoinducers was first observed in *V. harveyi* (Bassler et al., 1994). Besides the borated form of AI-2 described above (Surette et al., 1999), *V. harveyi* uses an AHL-type signal, HAI-1 (Cau and Meighen, 1989), as well as a signal of unknown structure, CAI-1 (Henke and Bassler, 2004). The inner membrane of *V. harveyi* contains cognate sensors for each signal and each of the three autoinducers presumably contains different information about the surrounding microbial community because AI-2 is produced by a multitude of bacteria, CAI-1 is produced by most sequenced vibrios, and HAI-1 is made exclusively by *V. harveyi* and a few closely related species. Two of the three signals are also used by *V. cholerae*, CAI-1 and AI-2. Remarkably, in both species, information from all the sensors feeds into a shared regulatory pathway (Fig. 2).

3.1.1 Signal Transduction via Phosphorelay

The autoinducer sensors of *V. harveyi* and *V. cholerae* belong to a large family of so-called two-component proteins, the same type of proteins many gram-positive bacteria use for sensing oligopeptide signals. In general, the sensor component of a two-component system acts as a kinase to phosphorylate the second component, a response regulator. The sensor's kinase activity depends on whether or not it is bound to signal. The response regulator typically binds DNA and acts as a transcriptional regulator, and its activity depends on whether or not it is phosphorylated.

When *V. harveyi* (and *V. cholerae*) is at low cell density and the sensors are not bound by autoinducers, the sensors act as kinases to phosphorylate the LuxU phosphotransfer protein, which in turn, phosphorylates the response regulator LuxO (Fig. 2). Phospho-LuxO (LuxO-P) indirectly represses synthesis of the global quorum-sensing regulator, LuxR (although they share the same name, *V. harveyi*'s LuxR is not homologous to the LuxR-type proteins described above). At high cell density, the autoinducers accumulate in the medium and bind their respective sensors. In response, the sensors change from acting as kinases to acting as phosphatases and remove phosphate from LuxO. Without phosphorylation, LuxO is inactive and repression of *luxR* is relieved. Thus, LuxR is produced and binds the promoters of quorum-sensing target genes, such as those needed for bioluminescence.

3.1.2 Multiple Small RNAs Act at the Core of the Quorum-Sensing Pathway

Repression of LuxR by LuxO was assumed to be indirect, because LuxO belongs to a family of transcriptional activators. Thus, it was hypothesized that LuxO activates transcription of a gene, which encodes a repressor of LuxR. A genetic screen to identify this putative repressor yielded the RNA chaperone Hfq (Lenz et al., 2004).

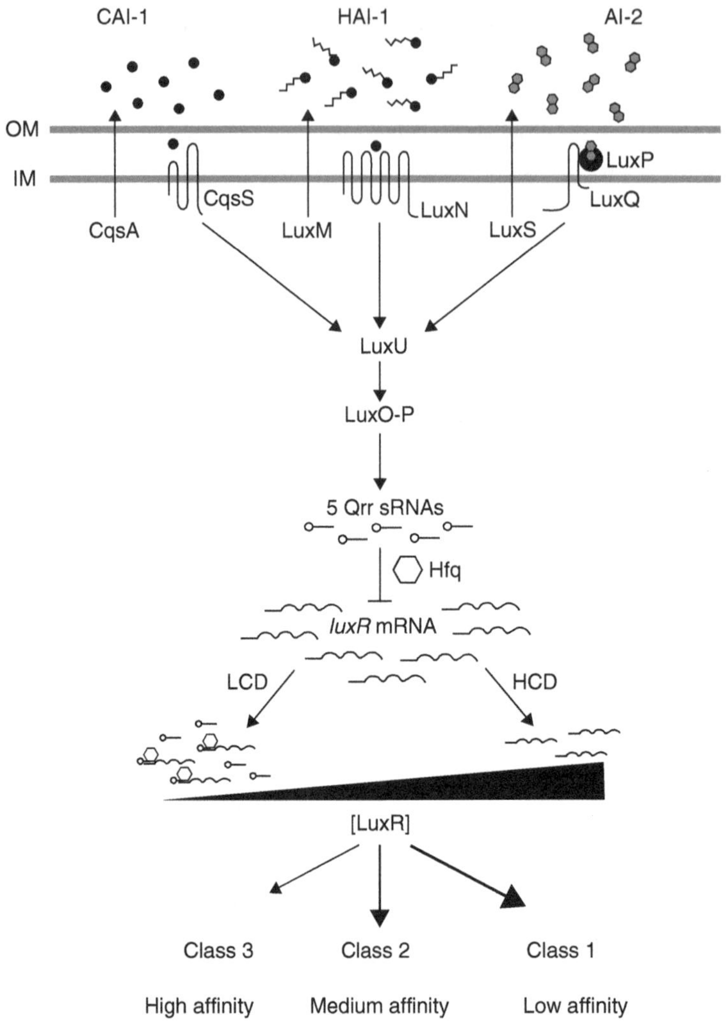

Fig. 2 Model of the quorum sensing pathway of *Vibrio harveyi*. Three parallel sensory systems converge to regulate quorum-sensing gene expression by controlling levels of the master transcriptional regulator, LuxR. The Qrr sRNAs (lollipops) and Hfq (hexagon) indirectly regulate LuxR protein levels by destabilizing *luxR* mRNA (wavy lines). The Qrr sRNAs produce an increasing gradient of LuxR protein as the cells transition from low to high cell density. See text for details. (OM) Outer membrane; (IM) Inner membrane; (LCD) Low cell density; (HCD) High cell density (Adapted from Tu and Bassler, 2007 and Waters and Bassler, 2006)

In recent years, Hfq has emerged as the key protein supporting the function of small regulatory RNAs (sRNAs). Hfq-dependent sRNAs are short (40–400 nucleotides), noncoding RNA molecules that function by basepairing with target mRNAs (Gottesman, 2004). This pairing typically blocks the ribosome binding site on the mRNA, thereby inhibiting its translation, or pairing can cause alterations in the secondary structure of the mRNA, which positively or negatively affect the translation efficiency and/or stability of the mRNA (reviewed in Storz et al., 2004; Gottesman, 2004). The role of Hfq in sRNA regulation is not fully understood. Hfq is known to bind both the sRNA and the target mRNA in many cases, and it is hypothesized that Hfq alters the secondary structure of either or both RNAs to increase pairing efficiency, or that Hfq simply brings the sRNA and its target mRNA in close proximity, increasing the local concentrations of the binding partners (Storz et al., 2004). sRNAs regulate the response to many environmental signals including iron, oxidative stress, elevated temperature, and toxic agents (Storz et al., 2004).

Finding Hfq in the vibrio screen implicated sRNAs in the quorum-sensing circuit of *V. harveyi* (as well as in *V. cholerae* where LuxO-P indirectly represses the LuxR homologue HapR). A bioinformatics screen of the sequenced *V. cholerae* genome revealed four genes encoding sRNA candidates that fit the screening criteria. These were named Qrr1–4 (for quorum regulatory RNA, Lenz et al., 2004). The bioinformatics approach could not be applied to *V. harveyi*, because the genome sequence was not available, but a combination of bioinformatics and genetic screening recently identified five Qrr sRNAs in *V. harveyi* (Tu and Bassler, 2007).

At low cell density, LuxO-P activates transcription of the Qrr sRNAs, and together with Hfq they destabilize the *luxR/hapR* mRNAs (Fig. 2: LCD, Lenz et al., 2004; Tu and Bassler, 2007). At high cell density the Qrr sRNAs are not produced because LuxO is inactive, so LuxR/HapR accumulate and regulate their target genes (Fig. 2: HCD). The Qrr sRNAs likely function by causing the coupled degradation of the sRNA:mRNA complex (Massé et al., 2003; Lenz et al., 2004). This mechanism provides an intrinsic means for terminating sRNA activity when the signals for their synthesis are no longer present.

3.1.3 Redundant Versus Additive Control

All nine Qrr sRNAs from *V. harveyi* and *V. cholerae* are clearly homologous; they have high sequence identity, and the secondary structures of the sRNAs are predicted to be nearly identical (Lenz et al., 2004; Tu and Bassler, 2007). However, the promoter regions of the *qrr* genes are not conserved besides the LuxO- and RNA polymerase sigma-factor binding sites. This suggests that the expression of each individual *qrr* gene could be subject to control by distinct regulators other than LuxO. Strikingly, although the components of the quorum-sensing pathway are almost identical between *V. harveyi* and *V. cholerae*, the individual contribution of each *qrr* to *luxR/hapR* repression is markedly different. In *V. cholerae*, the four Qrr sRNAs appear to be fully redundant. Any one of them is sufficient for full repression of HapR at low cell density, so the wildtype quorum-sensing response is only lost if all four *qrr* genes are simultaneously deleted (Lenz et al., 2004). By contrast, in *V. harveyi*, each of the Qrr sRNAs is capable of repressing *luxR* expression to a certain degree, but it requires four Qrr sRNAs to obtain full repression of *luxR* (the fifth *qrr* gene, *qrr*5, is not expressed under standard laboratory conditions)

(Tu and Bassler, 2007). It is difficult to account for the different mechanistic solutions evolved in *V. harveyi* and *V. cholerae*, but presumably each solution offers advantages unique to each species' lifestyle and environmental niche.

3.2 Signal Integration

3.2.1 Different Autoinducer Inputs Elicit Discrete Responses from Promoters in the Quorum Sensing Regulon of *Vibrio harveyi*

As described above, information from all three quorum-sensing signals in *V. harveyi* is transmitted via a phosphorelay to the central regulator LuxO, and ultimately through the Qrr sRNAs to the master regulator LuxR. To investigate whether this molecular arrangement allows distinct changes in gene expression in response to each autoinducer, Waters and Bassler (2006) examined the response of over 50 LuxR-regulated promoters to different combinations of HAI-1 and AI-2. The third autoinducer, CAI-1, is the weakest signal in *V. harveyi*, and it did not significantly affect the expression of any of the promoters examined. The strength of each AI-signal depends on the potency of the corresponding sensor's kinase- and phosphatase-activity (Freeman and Bassler, 1999; Freeman et al., 2000; Henke and Bassler, 2004).

The quorum-sensing regulated promoters fell into three classes: Class 1 displayed coincidence behavior; only when both HAI-1 and AI-2 were present simultaneously did the promoters elicit a significant response. Class 2 promoters showed an intermediate response to either autoinducer alone, and a full response to the simultaneous presence of HAI-1 and AI-2. Class 3 promoters responded fully to the presence of either or both autoinducers.

3.2.2 Different AI Inputs Translate into Distinct Levels of Qrr sRNAs and LuxR

The distinct responses of the quorum-sensing target genes to different combinations of HAI-1 and AI-2 can be traced back to different expression levels of the Qrr sRNAs under each autoinducer-condition. In the absence of autoinducers, each of *qrr*1–4 is fully expressed. In the presence of saturating levels of AI-2, the *qrr* genes are each expressed to a lesser extent. In the presence of saturating levels of HAI-1 *qrr* expression is further diminished. Finally, in the presence of saturating levels of both autoinducers *qrr* expression is almost eliminated. This pattern is presumed to be the result of distinct levels of phosphorylated LuxO being made in response to the each autoinducer input, but currently no technique exists to measure the phosphorylation level of a bacterial response regulator *in vivo*.

Importantly, in a qurom-sensing population individual cells display similar responses to the various autoinducer inputs (Waters and Bassler, 2006). Each level of Qrr sRNAs is translated into a discrete level of LuxR protein; the higher the Qrr sRNA level, the lower the LuxR level (Fig. 2). Preliminary evidence supports a model where Class 1, 2, and 3 promoters have discrete affinities for LuxR. Class 1 promoters have the lowest LuxR affinity, so they only respond to high concentrations of LuxR, consistent with the coincidence behavior, Class 2 promoters have a medium affinity for LuxR, and Class 3 promoters have the highest affinitiy for LuxR, explaining why they show a full response to either one of the autoinducers. Some aspects of this system are still not understood. Can *V. harveyi*, for example, distinguish between an intermediate level of HAI-1 and saturating levels of AI-2, or will these two conditions result in the same degree of phosphorylation of LuxO?

3.2.3 Additional Inputs into the Quorum Sensing Pathway of *V. cholerae*

Besides the two known autoinducers, at least two additional regulatory inputs feed into the quorum-sensing pathway of *V. cholerae*. First, the DNA-bending protein Fis binds the *qrr* promoters and increases LuxO-P-dependent activation (Fig. 3, Lenz and Bassler, 2007). *fis* expression is induced in early exponential phase and declines in stationary phase due to the *fis* promoter's sensitivity to CTP (Walker et al., 2004). Thus, Fis-mediated activation of the *qrr* promoters couples the quorum-sensing pathway with metabolic information about the growth phase of the cell.

Second, the VarS/VarA two-component system indirectly represses the level of LuxO phosphorylation (Fig. 3, Lenz et al., 2005). VarS kinase activity depends on an extracellular signal, which has not yet been identified. The signal has been suggested to accumulate during stationary phase, suggesting that VarS/VarA funnel information about the growth-phase of the population into the quorum-sensing pathway (Lenz et al., 2005). If so, the effect of both VarS/VarA- and Fis-mediated regulation is to repress entry into social mode when the population is growing exponentially, but increase entry into social mode when the population is in stationary phase.

3.3 The Qrr sRNAs Circumvent the Conventional Quorum Sensing Pathway

LuxR and HapR have been designated the master quorum sensing regulators in *V. harveyi* and *V. cholerae*, respectively, because all previously known target genes in the quorum sensing regulons depend on LuxR/HapR for their regulation. Accordingly, strains that carried mutations in LuxR or HapR were deemed incapable of quorum sensing. This concerns, among others, the classical strains of *V. cholerae*, which were responsible for previous pandemics of cholera (Joelsson et al., 2006). In a recent screen for additional members of the quorum sensing regulon, Hammer

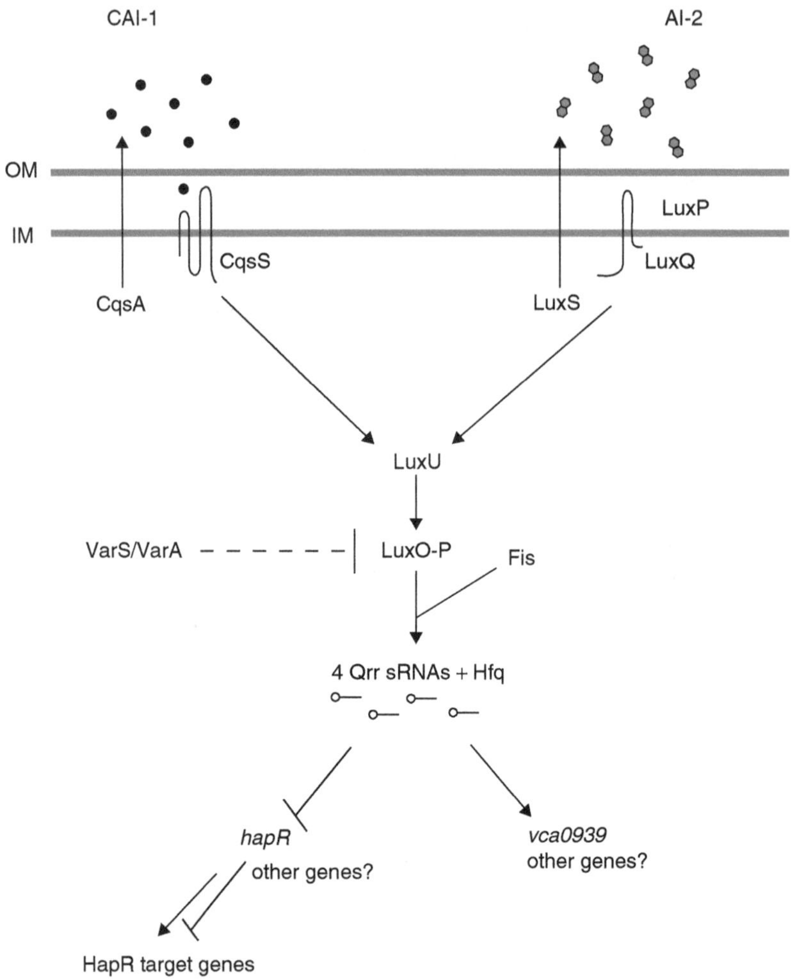

Fig. 3 Model of the quorum sensing pathways of *Vibrio cholerae*. Qrr sRNAs and Hfq control both HapR-dependent and HapR-independent quorum-sensing pathways. The flow of phosphate indicated by the arrows depict the low-cell-density state. The flow is reversed at high cell density. The dashed line indicates indirect repression at high cell density. See text for details. (OM) Outer membrane; (IM) Inner membrane (Adapted from Tu and Bassler, 2007 and Hammer and Bassler, 2007)

and Bassler (2007) discovered a new target gene, *vca0939*, which is regulated by quorum sensing but does not depend on HapR (Fig. 3). Instead, *vca0939* is directly regulated by the Qrr sRNAs. In the absence of sRNAs, *vca0939* translation is inhibited by a stem loop structure in the mRNA that occludes the ribosome binding site. Binding of a Qrr sRNA to the *vca0939* mRNA prevents formation of the inhibitory structure, allowing translation of the mRNA. *vca0939* was identified in *V. cholerae* El Tor, which has a functional *hapR* gene, but the authors went on to show that *vca0939* is also regulated by quorum sensing in classical strains of *V. cholerae*. Noticeably, all four *qrr* DNA sequences are 100% conserved in all sequenced strains of *V. cholerae*, suggesting that selective pressure exists to maintain the function of the Qrr sRNAs irrespective of the presence or absence of HapR (Hammer and Bassler, 2007). Thus, even strains that lack *hapR* are capable of cell-to-cell communication.

4 Conclusion

sRNAs play a critical role in the regulation of bacterial communication, not only in the vibrios discussed here, but also in distantly related species such as *P. aeruginosa, E. carotovora* and *S. aureus* (reviewed in Bejerano-Sagie and Xavier, 2007). It is not obvious why sRNAs would be employed for this role rather than proteins, but sRNA regulation is presumably advantageous when a rapid response is required because of the short time required to synthesize or degrade a sRNA compared to a protein. In the wild, bacteria presumably experience rapid changes in their cell densities as they move about, get incorporated into biofilms or enter/exit a host. It is likely critical for many of these species to rapidly switch into and out of quorum-sensing mode to adapt to these changes in cell density.

In the systems described here, multiple sRNAs act to control quorum sensing. Potentially, each sRNA could be regulated by different inputs, which would increase the plasticity of the quorum-sensing circuits. Moreover, the Qrr sRNAs of *V. cholerae* regulate at least one target in addition to *hapR*, and depending on the nature of the target, they can either repress or activate translation. In all likelihood, additional Qrr targets are yet to be discovered in both *V. cholerae* and *V. harveyi*. It is formally possible that the individual Qrr sRNAs each have separate targets, but their high degree of sequence identity make it implausible that an mRNA could be recognized by one but not another of the Qrrs.

To date, more than 80 small RNAs have been identified in the model bacterium *Escherichia coli* alone. Most of these control their targets at the mRNA level, but a few interfere with transcription (Storz et al., 2006) and some interact directly with target proteins (Gottesman, 2004). Thus, sRNAs are versatile regulators, they are involved in a diverse set of regulatory pathways and are found in significant numbers in many, if not all, bacterial species.

The list of quorum-sensing bacteria is rapidly growing, and so is our knowledge of the chemical vocabulary with which they communicate. Enticing examples of intraspecies, interspecies and even interkingdom communication have been described, but undoubtedly, since it is estimated that only a tiny fraction of Earth's bacteria have been identified, we are still far away from understanding the extent and significance of cell-to-cell communication in the bacterial world.

Acknowledgements I am grateful to Prof. Bonnie Bassler for critical reading of the manuscript and to Colleen O'Loughlin and Kimberly Tu for help with the figures.

References

Bassler BL and Losick R, 2006. Bacterially speaking. *Cell* 125:237–246

Bassler BL, Wright M, and Silverman MR, 1994. Multiple signaling systems controlling expression of luminescence in *Vibrio harveyi*: sequence and function of genes encoding a second sensory pathway. *Mol Microbiol* 13:273–286

Bejerano-Sagie M and Xavier KB, 2007. The role of small RNAs in quorum sensing. *Curr Opin Microbiol* 10:189–198

Cau J-G and Meighen EA, 1989. Purification and structural identification of an autoinducer for the luminescence system of *V. harveyi*. *J Biol Chem* 264:21670–21676

Chen X, Schauder S, Potier N, Van Dorsselaer A, Pelczer I, Bassler BL, and Hughson FM, 2002. Structural identification of a bacterial quorum-sensing signal containing boron. *Nature* 415: 545–549

Dong Y-H, Wang L-H, Xu J-L, Zhang H-B, Zhang X-F, and Zhang L-H, 2001. Quenching quorum-sensing-dependent bacterial infection by an *N*-acyl homoserine lactonase. *Nature* 411:813–817

Dong Y-H, Wang L-H, and Zhang L-H, 2007. Quorum-quenching microbial infections: mechanisms and implications. *Phil Trans R Soc B* 362:1201–1211

Duan K, Dammel C, Stein J, Rabin H, and Surette MG, 2003. Modulation of *Pseudomonas aeruginosa* gene expression by host microflora through interspecies communication. *Mol Micro* 50:1477–1491

Eberhard A, Burlingame AL, Eberhard C, Kenyon GL, Nealson KH, and Oppenheimer NJ, 1981. Structural identification of autoinducer of *Photobacterium fischeri* luciferase. *Biochemistry* 20:2444–2449

Freeman JA and Bassler BL, 1999. A genetic analysis of the function of LuxO, a two-component response regulator involved in quorum sensing in *Vibrio harveyi. Mol Micro* 31:665–677

Freeman JA, Lilley BN, and Bassler BL, 2000. A genetic analysis of the functions of LuxN: a two-component hybrid sensor kinase that regulates quorum sensing in *Vibrio harveyi. Mol Micro* 35:139–149

Gottesman S, 2004. The small RNA regulators of Escherichia coli: roles and mechanisms. *Annu Rev Microbiol* 58:303–328

Hammer BK and Bassler BL, 2007. Regulatory small RNAs circumvent the conventional quorum sensing pathway in pandemic *Vibrio cholerae. Proc Natl Acad Sci USA* 104(27):11145–11149.

Henke JM and Bassler BL, 2004. Three parallel quorum-sensing systems regulate gene expression in *Vibrio harveyi. J Bacteriol* 186:6902–6914

Hentzer M and Givskov M, 2003. Pharmacological inhibition of quorum sensing for the treatment of chronic bacterial infections. *J Clin Invest* 112:1300–1307

Joelsson A, Liu Z, and Zhu J, 2006. Genetic and phenotypic diversity of quorum-sensing systems in clinical and environmental isolates of *Vibrio cholerae. Infect Immun* 74:1141–1147

Lenz D and Bassler BL, 2007. The small nucleoid protein Fis is involved in *V. cholerae* quorum sensing. *Mol Micro* 63:859–871

Lenz D, Mok KC, Lilley BN, Kulkarni RV, Wingreen NS, and Bassler BL, 2004. The small RNA chaperone Hfq and multiple, small RNAs control quorum sensing in *Vibrio harveyi* and *Vibrio cholerae. Cell* 118:69–82

Lenz D, Miller MB, Zhu J, Kulkarni RV, and Bassler BL, 2005. CsrA and three redundant small RNAs regulate quorum sensing in *Vibrio cholerae. Mol Micro* 58:1186–1202

Lyon GJ, Wright JS, Christopoulos A, Novick RP, and Muir TW, 2002. Reversible and specific extracellular antagonism of receptor-histidine kinase signaling. *J Biol Chem* 277:6247–6253

Mashburn LM, and Whiteley M, 2005. Membrane vesicles traffic signals and facilitate group activities in a prokaryote. *Nature* 437:422–425

Massé E, Escorcia FE, and Gottesman S, 2003. Coupled degradation of a small regulatory RNA and its mRNA targets in Escherichia coli. *Genes Dev* 17:2374–2383

Mayville P, Ji G, Beavis R, Yang H, Goger M, Novick RP, and Muir TW, 1999. Structure-activity analysis of synthetic autoinducing thiolactone peptides from *Staphylocossus aureus* responsible for virulence. *Proc Natl Acad Sci USA* 96:1218–1223

McNab R, Ford SK, El-Sabaeny A, Barbieri B, Cook GS, and Lamont RJ, 2003. LuxS-based signaling in Streptococcus gordonii: autoinducer 2 controls carbohydrate metabolism and biofilm formation with Porphyromonas gingivalis. *J Bacteriol* 185:274–284

Miller ST, Xavier KB, Campagna SR, Taga ME, Semmelhack MF, Bassler BL, and Hughson FM, 2004. A novel form of the bacterial quorum sensing signal AI-2 recognized by the *Salmonella typhimurium* receptor LsrB. *Mol Cell* 15:677–687

Minogue TD, Wehland-von Trebra M, Bernhard F, and von Bodman SB, 2002. The autoregulatory role of EsaR, a quorum-sensing regulator in *Pantoea stewartii* ssp. *stewartii*: evidence for a repressor function. *Mol Micro* 44:1625–1635

Ohtani K, Hayashi H, and Shimizu T, 2002. The *luxS* gene is involved in cell-cell signaling for toxin production in *Clostridium perfringens. Mol Microbiol* 44:171–179

Osaki T, Hanawa T, Manzoku T, Fukuda M, Kawakami H, Suzuki H, Yamaguchi H, Yan X, Taguchi H, Kurata S, and Kamiya S, 2006. Mutation of *luxS* affects motility and infectivity of *Helicobacter pylori* in gastric mucosa of a Mongolian gerbil model. *J Med Microbiol* 2006, 55:1477–1485

Rickard AH, Palmer RJ, Blehert DC, Campagna SR, Semmelhack MF, Egland PG, Bassler BL, and Kolenbrander PE, 2006. Autoinducer 2: a concentration-dependent signal for mutualistic bacterial biofilm growth. *Mol Micro* 60:1446–1456

Schnauder S, Shokat K, Surette MG, and Bassler BL, 2001. The LuxS family of bacterial autoinducers: biosynthesis of a novel quorum-sensing signal molecule. *Mol Microbiol* 41:463–476

Storz G, Opdyke JA, and Zhang A, 2004. Controlling mRNA stability and translation with small, noncoding RNAs. *Curr Opin Microbiol* 7:140–144

Storz G, Opdyke JA, and Wassarman KM, 2006. Regulating bacterial transcription with small RNAs. *Cold Spring Harb Symp Quant Biol* 71:269–273

Surette MG, Miller MB, and Bassler BL, 1999. Quorum sensing in *Escherichia coli, Salmonella typhimurium*, and *Vibrio harveyi*: a new family of genes responsible for autoinducer production. *Proc Natl Acad Sci USA* 96:1639–1644

Tortosa P and Dubnau D, 1999. Competence for transformation: a matter of taste. *Curr Opin Microbiol* 2:588–592

Tu KC and Bassler BL, 2007. Multiple small RNAs act additively to integrate sensory information and control quorum sensing in *Vibrio harveyi. Genes Dev* 21:221–233

Visick KL, Foster J, Doino J, McFall-Ngai M, and Ruby EG, 2000. *Vibrio fischeri lux* genes play an important role in colonization and development of the host light organ. *J Bacteriol* 182:4578–4586

Walker KA, Mallik P, Pratt TS, and Osuna R, 2004. The *Escherichia coli fis* promoter is regulated by changes in the level of its transcription initiation nucleotide CTP. *J Biol Chem* 279:50818–50828

Waters CM and Bassler BL, 2005. Quorum sensing: cell-to-cell communication in bacteria. *Annu Rev Cell Dev Biol* 21:319–346

Waters CM and Bassler BL, 2006. The *Vibrio harveyi* quorum-sensing system uses shared regulatory components to discriminate between multiple autoinducers. *Genes Dev* 20:2754–2767

Whitehead NA, Barnard AML, Slater H, Simpson NJL, and Salmond GPC, 2001. Quorum-sensing in Gram-negative bacteria. *FEMS Microbiology Reviews* 25:365–404

Wu H, Song Z, Hentzer M, Andersen JB, Molin S, Givskov M, and Hoejby N, 2004. Synthetic furanones inhibit quorum sensing and enhance bacterial clearance in *Pseudomonas aeruginosa* lung infection in mice. *J Antimicrob Chemother* 53:1054–1061

Xavier KB and Bassler BL, 2003. LuxS quorum sensing: more than just a numbers game. *Curr Opin Microbiol* 6:191–197

Dynamical Genetic Regulation

Mogens H. Jensen, Sandeep Krishna, Kim Sneppen, and Guido Tiana

Abstract The development of new techniques to investigate quantitatively the expression of genes in the cell has shed light on a number of systems which display regular oscillations in protein concentration. In order to investigate the physics which controls such systems, one can make use of models based on rate equations. The models present in the literature suggest that important ingredients to describe correctly oscillations in protein expression are a feed–back loop, time delay and saturated binding. In this review we will discuss the physical aspects of the mechanism which is at the basis of the oscillations in gene expression.

Keywords Oscillations, feed-back loops.

1 Introduction

The standard view of biology has been for many years that the regulation of gene expression is a process which takes place at chemical equilibrium. In this framework, the cellular response to a stimulus cannot but be a shift in the equilibrium concentration of proteins or RNA. For instance, a damage in DNA produces kinases which phosphorylate p53; phosphorylation decreases the *equilibrium* binding constant to Mdm2, a protein which favours the degradation of p53; the *equilibrium*

Mogens H. Jensen(✉), Sandeep Krishna, and Kim Sneppen
Niels Bohr Institute
Blegdamsvej 17, DK-2100 Copenhagen Ø
Denmark
mhjensen@nbi.dk

Guido Tiana
Department of Physics
University of Milano and INFN via Celoria 16
20133 Milano
Italy

concentration of p53 increases. In this picture, production and degradation of molecules are always balanced.

The historical rationale which lies behind this picture is that the experimental tools used to investigate protein expression have been, for a long time, static. Western blots, the most common tool for this purpose, give a snapshot of the cell at the moment when the cell is lysated. It is quite impractical to describe the kinetics of the concentration of a protein with this technique because of a number of problems, like the synchronization of the cells at the initial time, the fact that cells need to be lysated to get a measurement and the qualitative character of the result. Nonetheless, in 1997 the group of Moshe Oren showed by damaging the DNA of different cell samples by UV–light and lysating them at intervals of 30 min, that the concentration of p53 in the cell displays periodic time oscillations [1], something very different from the static picture accepted sofar.

The development of fluorescence techniques made the study of the kinetics of gene expression within the cell much easier. Inserting the gene of a fluorescent protein in the DNA of a cell allows to measure in a quantitatively careful way the coexpression of any given protein. Using several proteins, fluorescent at different wavelengths, allow to study even the correlation between the expression of different proteins. This kind of experiment highlighted, in recent years, that protein expression can undergo complicated kinetics, the most outstanding of which is sustained time oscillations.

Periodic time oscillations have been observed at least in three cellular systems: the expression of p53 regulated by Mdm2 [1], the expression of Hes1 regulated by its own mRNA [2] and the expression of NFκB regulated by IkB [3]. As shown in Fig. 1, oscillations in protein concentration last for the whole duration of the experiments without appreciable decrease of the height of the peaks, suggesting that no or little dumping mechanism is involved.

All these oscillating systems involve two proteins, in the sense that, although these two proteins can interact with many others, knocking down these others does not change the oscillating character of the system. Consequently one can infer that two proteins are enough to produce oscillations. Moreover, these two proteins build out, in all cases, a negative feed–back loop where a protein and its repressor display time oscillations, either dumped or sustained, as a response to an external stimulus.

Another kind of time oscillations in biological systems are that in circadian clocks. Since they rely on a different physical mechanism, which has been widely studied elsewhere [4], we will not deal with them here.

The biological aim of these oscillations, if any, has not been fully understood. Most likely, it is strictly connected with the function of each specific protein. One should anyway remember that oscillating waves are the most widespread way of sending signals in physics and biology. This because an oscillating signal can carry a much richer set of information (its period, its phase, etc.) than a mere intensity shift. However, in this review we are not much interested in the consequences of oscillations in gene expression, rather than in the physical mechanism which produce them. In particular, we suggest that the basic ingredients which produces oscillations are the same for the three expression systems described above, and are feed–back loops, time delay and saturated binding.

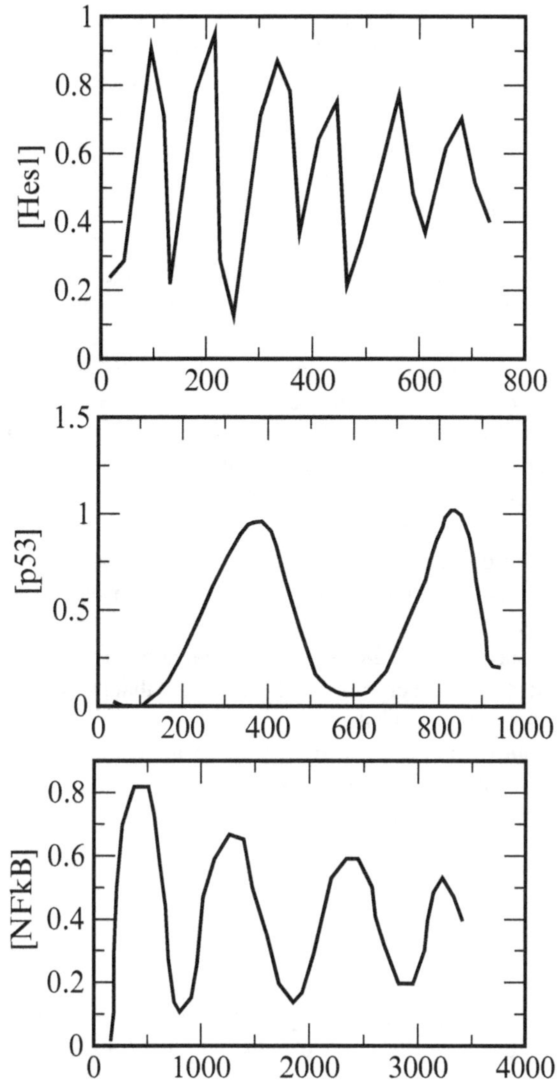

Fig. 1 The experimentally observed oscillations in Hes1 (upper panel, data taken from [2]), p53 (middle panel, data taken from [15]) and NFκB (lower panel, data taken from [3])

In order to reach this conclusion we need a model to describe the biological system, that is a coarse–grained description of the system, based on a limited number of ingredients and which reproduces the experimental results in which we are interested. In order to avoid conceptual mistakes, we think it is important to emphasize some facts concerning biological modelling. First, one should search for the simplest description (in terms of number of equations, parameters, etc.) which reproduces the relevant experimental data. Only in this way one can hope to understand (i.e., not just to describe) the underlying physical mechanism. Also, essentially any model with a large–enough number of parameters can fit any set of experimental data, even if the physics described by this oversized model is wrong. Aside from not giving any insight in the physical mechanism, such over–fitted model will hardly be robust with respect to small changes. Another important issue to remember is that, in general, the range of validity of a simplified model is quite limited. In other words, a model is designed to describe some aspects of a biological system, under well-defined conditions, to reproduce some experimental results and some small variations of them; extending the model to other aspects of the system or changing consistently the conditions usually leads to catastrophic outcomes. The risk that this happens is larger, the larger is the number of parameters and equations.

The model we will employ to account for feed–back loops describes the cell in terms of the mean concentration of molecules (proteins or RNA). The state of a cell is thus characterized by the values of the mean concentration of the molecules of interest. We will then disregard all the aspects of the kinetics connected with diffusion and translocation, assuming that these processes are much faster than the change in molecule concentration (cf. Table 1), so that they are always in a stationary state. When one wishes to differentiate between nuclear concentration and cytoplasmic concentration, such as in some models of NF-κB, one can account for them as two different molecules (see Sect. 2.4).

The molecule concentrations are regarded as real numbers. This is an approximation which can fail at low concentration, where the fact that the number of molecules is a discrete quantity can play a role.

Under these approximations, the dynamics of molecule concentrations is described by a system of rate equations, that is, for the case of two species X and Y, whose concentrations are x and y,

$$\begin{aligned}\frac{dx}{dt} &= f(x,y) - k_x x \\ \frac{dy}{dt} &= g(x,y) - k_y y,\end{aligned} \tag{1}$$

where k_x and k_y are the spontaneous degradation rates of the two molecules. The functions f and g describe the molecular signalling which controls the kinetics.

Table 1 Time scales of some cellular processes associated with a single molecule. The upper part of the table indicate the processes which are usually neglected in writing the rate equations of regulation newtworks, while the lower part indicate processes which are usually accounted for

Translocation through nuclear pores [20]	10^{-4} s
Diffusion in eucaryotic cell	1 s
Translation [19]	30 s
Transcription [19]	3 min
mRNA degradation [19]	3 min
Protein degradation	10 min to 10 h
Cellular signals [1–3]	1h

Since signalling typically takes place as binding between two species, the functions are usually monotonic and of definite sign. For example, if Y is a transcription factor which activates X, then $f(x,y)$ will be a sigmoidal positive function of y.

More precisely, if the protein X (whose total concentration is x) binds to the operator O on the DNA (whose concentration o is usually constant), under the equilibrium conditions described above one can define the dissociation constant $k = [ox]/(o_{free} \cdot x_{free})$, where $[ox]$ is the concentration of X complexed to the operator O, $x_{free} \equiv x - [ox]$ and $o_{free} \equiv o - [ox]$. Under the further assumption that $[ox] \ll x$ then the fraction of filled operators is

$$\frac{[op]}{o} = \frac{x}{k^{-1} + x}, \tag{2}$$

which has a sigmoidal shape. This is the typical function used in Eq. (1), but of course not the only one. A common modification is to add an exponent h to each quantity in the right–hand side of Eq. (2) in order to take into account some cooperativity in the binding. The resulting equation is $[op]/o = x^h/(k^{-h} + x^h)$, and the exponent h, which is called Hill's coefficient, indicate the (effective) number of molecules which has to bind at the same time to obtain the desired effect. For example, in the case that the a transcription factor has to be in tetrameric form to bind to the operator, one would obtain $h = 4$.

Another important hypothesis which lies behind Eq. (1) is that changes in the concentration of a species cause their effects instantaneously. For example, if y increases, then the time derivative of x changes at the same very moment. This is certainly an approximation, because the function $f(x,y)$ hides a number of molecular processes which need some time to be carried out. The approximation done in writing Eq. (1) is that these processes are much faster than the velocity at which x and y vary.

1.1 Negative Feed–Back Loops

All these oscillating systems display a core of two molecules building out a feed–back loop, that is molecule X activates molecule Y while molecule Y represses molecule X. Feed–back loops are a particularly common motif in regulatory networks [5], because they can act as switches. An efficient switch in a regulatory network is a motif where the concentration of a protein displays a sharp variation when one of the parameters of the system, such as a binding constant or some rate of expression/degradation, overcomes a given threshold, while it is barely sensitive to the other parameters.

Of course, another motif which, in principle, produces changes in protein concentration is just a chain of nodes where the change in the parameters cause a change in the thermodynamic equilibrium of the system, ending up into a larger concentration of a given protein. For example, if the dissociation constant between a protein X and another protein Y which promotes ubiquitination of X increases, the concentration of X will increase. But this is not a switch, because the variation in the concentration of proteins is linear with respect to all the parameters which control it. It is like if the intensity of the lamp in your living room were proportional to the pressure of your finger on every button of your home. Quite impractical.

A feed–back loop made out of two nodes, each of which interacts instantaneously and does not feed itself, cannot produce steady oscillations. In fact, the most general model for the dynamics of two molecules (either protein or RNA), whose concentrations are x and y, is that of Eq. (1). If $\partial f/\partial x \leq 0$ and $\partial g/\partial y \leq 0$, as guaranteed by the hypothesis that each molecule does not feed itself, Dulac's criterion [6] excludes the possibility of closed orbits, and thus of sustained oscillations.

Note that this result is true only for systems built out of two molecules. Negative feed–back loops composed of more molecules interacting instantaneously can display steady oscillations. It is interesting to stress that a negative feed–back is not optional to have such oscillations. In fact, a conjecture by R. Thomas [7], later proved by Gouzé [8], states that the presence of a negative loop in the network is a necessary condition for stable periodic behaviour.

Moreover, we are investigating only the conditions which allow sustained oscillations, because experimental data suggest that dumping does not occur, or occurs marginally. Dumped oscillations are easier to obtain even in a two–species feed–back loop. However these oscillations are not robust with respect to the initial concentrations of the species, and consequently their possible biological role are uninteresting.

1.2 Time Delay

The impossibility of a two-species loop to produce oscillations can be easily solved abandoning the unrealistic hypothesis that the interactions between the different molecules in Eq. (1) are instantaneous. In Table 1 are reported the time scales of some cellular processes which are hidden behind the functions which control the rate equations. If the variations in protein concentration described by Eq. (1) are much faster than diffusion, translation, transcription, etc., then it is safe to assume that the interactions are instantaneous and Eq. (1) is correct. Unfortunately this is seldom the case. For example, the oscillations observed in the protein concentration in [1–3] display a period of the order of one hour. On the other hand, protein expression and protein degradation takes place on the time scale of tens of minutes. It is consequently unrealistic to describe the associated feedback–loop with Eq. (1).

One can describe more realistically the kinetics of protein concentration by delay rate equations of the kind

$$\begin{aligned}\frac{dx(t)}{dt} &= f(x(t),y(t),x(t-\tau_1),y(t-\tau_2)) - k_x x(t)\\ \frac{dy(t)}{dt} &= g(x(t),y(t),x(t-\tau_3),y(t-\tau_4)) - k_y y(t),\end{aligned} \tag{3}$$

where the delays τ_1, τ_2, τ_3 and τ_4 summarize the time spent by the molecular processes not described explicitly. Of course this is the most general form, and often it is possible to take into account only fewer than four delays.

The effect of delay in a rate equation is typically that of causing oscillation. Even in the simplest case of a linear differential equation with delay

$$\frac{dx(t)}{dt} = -x(t-\tau) \tag{4}$$

the variable $x(t)$ displays oscillations if τ is large enough. The physics which produces such oscillations can be readily understood with an example: consider a person who walks along a straight line to a given point, marked on the ground. If this person is able to take instantaneous decisions, he will approach the mark and then stop. This is a stationary solution to the walk kinetics. If, on the other hand, it takes some time to realize that the mark has been reached, the person will not stop at the mark, but cross it. When eventually the information that the mark has been crossed is elaborated, the person will turn back and walk on the opposite direction. The mark will again be reached and overcome, and so on. The resulting kinetics will be sustained oscillations about the mark.

Of course, a linear delay rate equation is an oversimplification of an expression system, but the physics which lies behind delay rate equations like those of Eq. (3) is the same. Here there is the further complication that the functions f and g are in general strongly non–linear, resulting in an amplification of the effect of the delay.

An explicit delay in the rate equations is not the only way to account for time–consuming molecular mechanisms. Another way is to introduce auxiliary variables which vary slowly on time. For example, in Eq. (3) one could substitute the dependence of f on – say $-y(t-\tau_2)$ with a dependence on a tailor–made variable $z(t)$ which satisfies some kinetic equations whose solution is approximately equal to $z(t) = y(t-\tau)$. Note that the kinetic equation for $z(t)$ in general will have a solution similar to $z(t) = y(t-\tau)$ only in a limited range of variations of $y(t-\tau)$ (i.e., the range where one has fitted it), and will be hardly useful outside this range.

This kind of approach is that used by Bar Or et al. in [9] to describe the oscillations of p53. They introduce an unknown protein to provide the delay necessary to have oscillations. Whatever this protein is real or just a mathematical *escamotage* to summarize a number of events (e.g., transcription, translation) which are quantitatively not under control, this kind of approach is quite dangerous. First, because it certainly helps to fit the experimental data (increasing the number of parameters is always possible to fit anything) but it does not help to understand the physics which controls the system. Moreover, since this unknown protein has never been observed, it probably overfits the data, that is gives a description of the system which cannot be generalized in any direction. Changing even slightly the parameters of the system will produce unrealistic results, because they will depend on the details of the (unrealistic) kinetics of the tailor–made variable.

There are, on the other hand, cases where an implicit introduction of a delay by means of some auxiliary protein is useful. This is the case, as will be described in the next Section, of NFκB. In this case the delay is produced by a well—defined, single mechanism, that is the NFκB-dependent production of IkB mRNA. This mechanism can be easily described explicitly and the associated numerical parameters are known. As a rule of thumb, the delaying mechanism should be made explicit only when it is well-defined and its parameters are under control.

1.3 Other Issues

Another common feature of the three negative loops found to display sustained oscillations is that the binding of the two species is saturated, in the sense that their dissociation constant k_D is much larger than their concentration. For example, in the case of p53–Mdm2 k_D is of the order of 500 nM [10], while in the case of NFκB–IkB is 1 nM [3] (while the typical concentration of these proteins in the cell is of the order of 10 μM). This means that the vast majority of them is bound together, or equivalently, that the amount of free protein is small. As will be discussed in more detail in Sect. 2, dealing with specific examples, the saturated character of the binding is crucial to produce oscillations.

Another issue is whether these systems can display more complex dynamical behavior other than a stable fixed point or oscillations, most interestingly if they are capable chaotic dynamics, at least for some values of the parameters. It is well known [6] that chaotic behavior is rather common, even in simple, low–dimensional dynamical systems. If the systems we are studying can display chaotic dynamics, then a related problem is how evolution could deal with this region of the parameters space. Indeed, it is not obvious if the cell can somehow take profit of having a chaotic mechanism or, conversely, if chaos is deleterious. In the latter case, there should be a clear evolutionary pressure in these systems to stay away from the chaotic regimes.

Interestingly, general mathematical results have been derived for monotonic loops [11, 12]. Even in the presence of time delays, a monotone loop always behaves like a two-dimensional system, regardless of its length, and this means that it cannot be chaotic. Loosely speaking, after a transient one can only approach a stable periodic orbit or a fixed point. In some sense, these results support the view that feedback loops are good candidates as biological clocks, having a dynamics which is very robust qualitatively to changes in the parameter values. On the other hands, several studies (see, e.g., [13]) have shown how models with several intertwined feedback loops commonly have a richer range of dynamical regimes, including chaotic ones.

2 Oscillations in Biological Systems

Let alone circadian clocks, the negative feed–back loops built out of p53–Mdm2, Hes1–mRNA and NFκB–IkB are the only systems discovered up to date to display sustained oscillations. In the following, we will briefly give a detailed description of them, discussing their peculiarities but emphasizing that the basic physical mechanism which produces oscillations is the same.

2.1 p53-Mdm2

The protein p53 is responsible for inducing apoptosis in cells with damaged DNA [14]. The concentration of p53 is usually kept low by a feedback mechanism involving another protein, Mdm2, which binds to p53 and promotes its degradation. When the DNA is damaged, the cell expresses a number of kinases which phosphorylate SER20 in p53, changing its affinity to Mdm2. This results in oscillations in the concentration of p53, observed both in western blot analysis [1, 9] (cf. Fig. 1a) and in experiments on single cells performed coexpressing fluorescent proteins [15]. The accepted explanation for the overall increase in the concentration of p53 is that its phoshoprylation decreases its affinity to Mdm2, shifting the thermodynamical equilibrium towards higher concentrations.

Aside for not justifying the time oscillations, this explanation does not agree with a number of experimental evidences. Equilibrium isothermal titration calorimetry experiments have shown [10] that the effect of phosphorylation at SER20 is not to increase, but to decrease the dissociation constant between p53 and Mdm2 from $k_D = 575 \pm 19nM$ to $k_D = 360 \pm 3nM$. The same effect is observed *in vivo* [16], where p53ASP20 (a mutated form which mimics phosphorylated p53) binds Mdm2 more tightly than p53ALA20 (which mimics unphosphorylated p53). Moreover, single–cell experiments display [15] a slight decrease in the concentration of p53 after DNA damage, which cannot be explained by the classical theory.

It was shown in [17] that taking into account the time delay associated with some lengthy processes within the cell can account in a simple and physically clear way for the experimental facts. The loopback mechanism is sketched in Fig. 2a and the associated time–delayed rate equations are

$$
\begin{aligned}
\frac{\partial p}{\partial t} &= S - a\cdot pm - b\cdot p \\
\frac{\partial m}{\partial t} &= c\frac{p(t-\tau) - pm(t-\tau)}{k_g + p(t-\tau) - pm(t-\tau)} - d\cdot m \\
pm &= \frac{1}{2}\left((p+m+k) - \sqrt{(p+m+k)^2 - 4p\cdot m}\right),
\end{aligned}
\tag{5}
$$

where the delay τ takes into account the half–life of mRNA, the diffusion time, the time needed to cross the nuclear membrane and the transcription/translation time. One can solve numerically Eq. (5), simulating the damage in DNA with a sudden change in the dissociation constant k. Making use for the rates of the numerical values discussed in ref. [17] and changing instantaneously at time $t = 2000\,s$ the

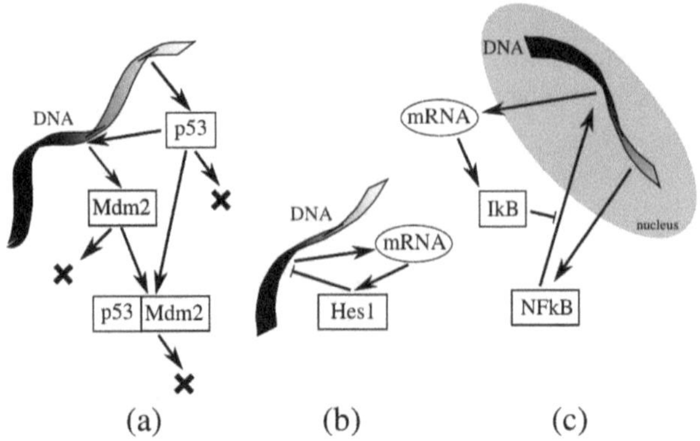

Fig. 2 A sketch of the feed–back loops mechanism controlling the concentration of p53 (a), Hes1 (b) and NKkB (c)

value of k from 180 to 18, one obtains the oscillations displayed in Fig. 3. A number of important facts emerge from these calculations. Most interestingly, what triggers oscillations is a decrease in the dissociation constant k, while any increase in its value just shifts the equilibrium towards higher values of p, without displaying and oscillation (cf. inset of Fig. 3). Moreover, just after $t = 2000\,s$ the concentration of p53 decreases, as observed in the experiments. Finally, the concentration pm of the complex p53–Mdm2 is at any time essentially identical to the minimum between p and m (gray curve in Fig. 3), indicating that the degradation is saturated.

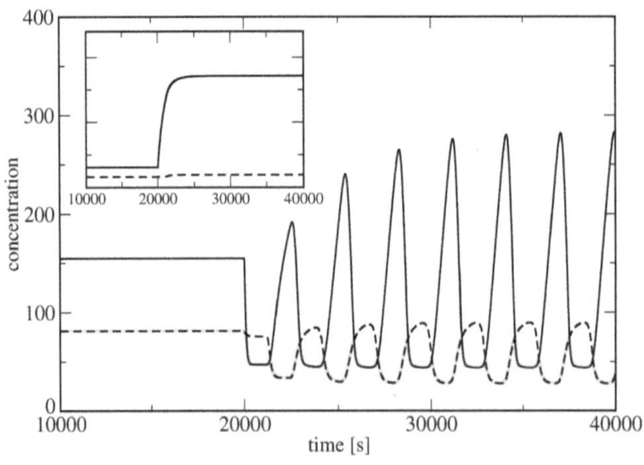

Fig. 3 Oscillations displayed by the numerical solution of the dynamic equations of p53. The numerical values of the rates are $a = 3 \cdot 10^{-2}\,s^{-1}$, $b = 10^{-4}\,s^{-1}$, $c = 1\,s^{-1}$, $d = 1\,s^{-2}$, $S = 1\,s^{-1}$, $k = 180$, $k_g = 28$ and $\tau = 1200\,s$. At time $t = 20000\,$s the value of k is decreased of a factor 10. In the inset, the solution of the same equations, where the value of k is increased of a factor 10 at $t = 20000\,$s. The equations are solved with the Adams algorithm

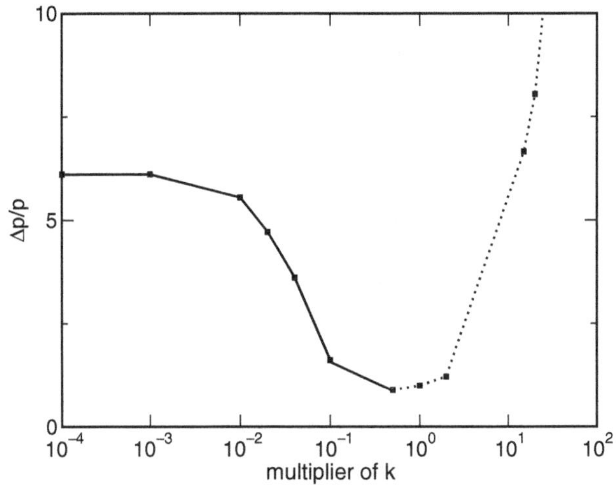

Fig. 4 The height of the response peak Δp with respect to the quantity that multiplies k, mimicing the stress. The dotted line indicates that the system does not display oscillatory behaviour

The relative change Δp in the maximum concentration of p53 reached after the simulated stress event is displayed in Fig. 4 as a function of the change in k which simulates the stress. The value of Δp displays a sigmoidal behaviour if the stress decreases the value of k, behaviour which is suited for a molecular switch. On the contrary, the steady state Δp which the system would experience without delay is linear with respect to k [17], thus differing markedly from a switch behaviour.

A detailed analysis of the response of the oscillations to changes in the parameters which define the system is also done in ref. [17]. It turns out that the oscillations do not change qualitatively for changes of the parameters a, b and c within five orders of magnitudes, while a decrease in d or k_g suppresses the oscillating behaviour. Since the failure of the p53 response mechanism is associated with a large number of tumors, one could speculate that the weak points to be further investigated to understand the uprise of such tumors are the degradation of Mdm2 and the binding of p53 to the DNA. As a matter of fact, around the 45% of all tumors display mutations in the p53 region which binds to DNA [21].

2.2 *Stability Analysis of p53-Mdm2*

The overall behaviour of the proteins with respect to time depends obviously on the parameters of the dynamic equations, that is the production and degradation rates and the delay. The long-time behaviours that the protein can display are either the convergence to a steady state or a limit cycle, which causes sustained oscillations in the concentration of the proteins. Chaotic behaviour is never observed within this model.

A careful analysis of how the dynamics of the protein concentration depends on the parameters of the rate equations has been done by Neamtu and coworkers in [18]. The main conclusion of this work is that under the condition that the dissociation constant k between p53 and mdm2 is small, there exist a critical delay τ_0 above which the system displays oscillations. In the language of dynamical systems, this is a Hopf bifurcation.

A Hopf bifurcation is the generic type of bifurcation that occurs when a stationary fixed point of the system becomes unstable and turns into a dynamically oscillating state on a limit cycle, as consequence of a change in some parameter of the system. In other words, a biological switch is, in the language of dynamical systems, a Hopf bifurcation.

2.3 Hes1 and ITS mRNA

The transcription factor Hes1 controls the differentiation of neurons in mammal embryos [22]. Its concentration is controlled by a feedback loop built out of Hes1 and its own mRNA (see Fig. 2). Like p53, its concentration display time oscillations if the cells are stimulated with serum [2]. The oscillation period is similar to the of p53, that is approximately two hours, and lasts for $\approx$ 12 hours. The aim of these time oscillations are not entirely clear, but are anyway likely to be connected with the creation of spatial patterns. In this respect, the timing is essential. In fact, the role of Hes1 is to repress another protein called Mash1; both the knocking out of Mash1 and its continuous expression by means of retroviral introduction results in a lack of cell differentiation. Only the time oscillations in the concentration of Hes1 (and thus of Mash1) guarantee a proper differentiation of neuronal cells [22], showing that time variations are critical for the development of the nervous system. The control mechanism again involves an activation and an inhibition: the transcription of the mRNA of Hes1 activates the translation of the protein, following the standard cellular mechanism, and Hes1 represses the transcription of its own mRNA. The main difference with respect to p53 is that here the repression is not carried out by another protein, but by the mRNA of Hes1 itself. Note however that from the physical point of view the difference is merley semantical: the control network still has two nodes, of which one is activating and one is repressing. The molecular species of these nodes are immaterial to the description of the oscillations. In the case of p53, we did not describe explicitly the mRNA of the two proteins involved because they did not change the nature (activating/repressing) of the associated node. Their effect was only to delay the expression of the protein when the transcription factors activates it. In a picture of the network made of nodes and arrows, the transcription and translation of p53 mRNA are just two consecutive arrows, that can be summed up in a single one if one takes into account the time needed for these two molecular processes with a proper delay. Of course, in the case of Hes1 and of all systems where a protein inhibits its own translation, the role of mRNA is much more important and needs to be consid.

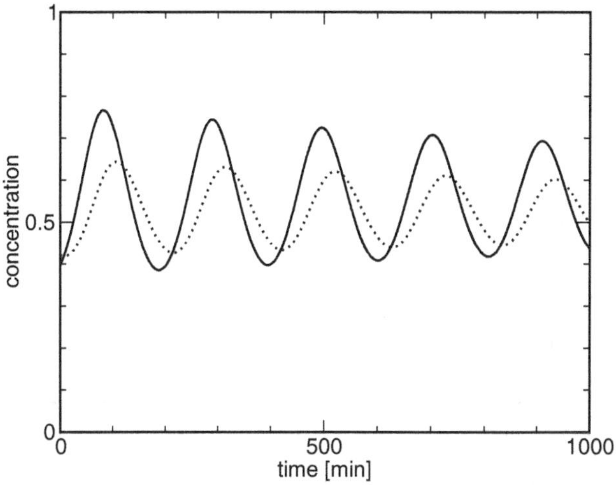

Fig. 5 The concentration of Hes1 (solid curve) and uts mRNS (dashed curve) obtained solving numerically Eq. (7)

From a physical point of view, the system can be described (cf. [23]) as

$$\frac{\partial r}{\partial t} = \frac{\alpha k^h}{k^h + [s(t-\tau)]^h} - k_r r(t)$$
$$\frac{\partial s}{\partial t} = \beta[r(t)] - k_s s(t), \tag{6}$$

where s and r are the concentrations of Hes1 and its mRNA, respectively. The meaning of these equations is that mRNA is produced at rate α when Hes1 is bound to the DNA. The probability that Hes1 is bound to DNA is $k^h/(k^h + s^h)$, where k is a characteristic concentration for dissociation of Hes1 from the DNA, and h is the Hill coefficient that takes into account the cooperative character of the binding process due to the fact that Hes1 is a dimer. k_r and k_s are the spontaneous degradation rates of the two proteins, while τ is the delay associated with the molecular processes that we do not want to describe explicitly (transcription, translocation, etc.). Ref. [2] suggests that τ_{rna} and τ_{hes1} are of the order of 25 min. The value of the time delay is difficult to assess, since it is determined by a variety of molecular processes. One can guess that its order of magnitude is tens of minutes. The numerical solution of Eq. (7) is displayed in Fig. 5. The oscillations have a period $\Delta\tau \approx 170$ min, while their dependence on the delay τ is displayed in Fig. 6. For any delay in the range $10 < \tau < 50$ min, the oscillation period is consistent with that found experimentally, and also the time difference between the peaks in Hes1 and mRNA is 18 min, similar to the experimental findings. For $\tau < 10$ min, the system shows no oscillations. To check the robustness of the results, we have varied α, β and k over 5 orders of magnitude around the basal values listed in the caption to Fig. 5, and observed no qualitative difference with the oscillatory behaviour described above. On the other hand, a decrease of k_s and k_r disrupts the oscillatory mechanism. This

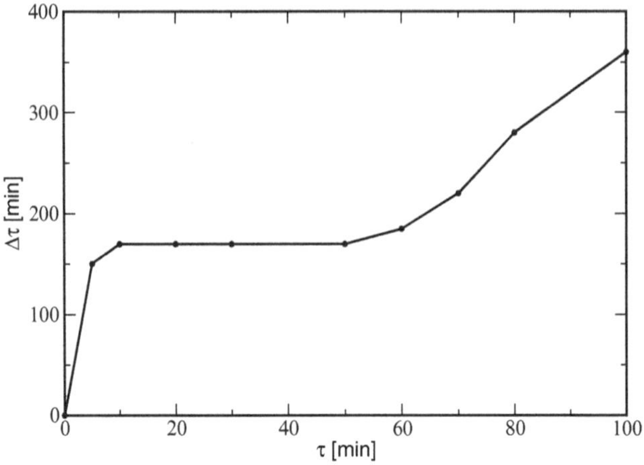

Fig. 6 The oscillation period $\Delta\tau$ of Hes1 as a function of the delay τ

is because these two quantities set the time scale of the system, with which τ has to be compared. Increasing such time scales at constant τ is equivalent to decreasing τ for a given time scale, and thus no oscillations are detected.

The behaviour of Hes1 is thus very similar to that of p53, both in the features of the time oscillations and in the lag delay before observing them. This is not unexpected, since the structure of Eq. (6) is very similar to the structure of Eq. (5).

2.4 NFκB and IκB

The NFκB family of proteins is one of the most studied in the last ten years, being involved in a variety of cellular processes including immune response, inflammation, and development. NF-κB can be activated by a number of external stimuli [24] including bacteria, viruses and various stresses and proteins (e.g., tumor necrosis factor-α, TNF-α, which was the signal used in [3, 25]). In response to these signals it targets over 150 genes including many chemokines, immunoreceptors, stress response genes, as well as acute phase inflammation response proteins [24].

Each NF-κB has a partner inhibitor called IκB, which inactivates NF-κB by sequestering it both in the nucleus as well as in the cytoplasm. In fact, the IκB proteins come in several isoforms $\alpha, \beta, \varepsilon$. Some of these isoforms are, in turn, transcriptionally activated by NF-κB, thus forming a negative feedback loop which is essentially identical in structure to the other two discussed above.

The potential for this negative feedback loop to produce oscillations in the nuclear-cytoplasmic translocation of NF-κB was initially shown by the electrophoretic mobility shift assay experiments of [3]. They found that wild-type cells show some damped oscillations, while mutants containing only the IκBα isoform showed sustained oscillations. In contrast, cells with only the IκBβ or ε isoforms do not show oscillations. This conclusion was bolstered by single-cell fluorescence imaging experiments which show sustained oscillations NF-κB transcription factor in mammalian cells [25], with a time period of the order of hours. In these experiments, IκBα was overexpressed, hence the system behaves like the mutant which only has the IκBα isoform.

This mutant, referred to as the $\beta -/-; \varepsilon -/-$ mutant, is the cleanest case to model sustained oscillations involving NF-κB, therefore we will focus our modelling on this mutant. The following cellular processes, summarized in Fig. 2c, are important for this system on the timescales we are interested in (i.e., we ignore processes which are very slow):

- NF-κB, when in the nucleus, activates transcription of the IκBα gene (henceforth we will drop α unless we are explicitly talking about more than one isoform), producing IκB mRNA in the cytoplasm.
- The IκB mRNA is translated to form IκB protein.
- The IκB protein can be transported in and out of the nucleus.
- In both compartments, IκB forms a complex with NF-κB.
- The NF-κB-IκB complex (henceforth referred to as $\{$NI$\}$) cannot be imported into the nucleus. However, if it forms within the nucleus it can be exported out.
- Free NF-κB behaves in exactly the opposite way. Free NF-κB is actively transported into the nucleus but not from the nucleus to the cytoplasm.
- The cytoplasmic $\{$NI$\}$ complex is tagged by another protein, the IκB kinase (IKK), for proteolytic degradation. This results in degradation of IκB only, releasing NF-κB. Note that this degradation does not occur for free IκB.

Notice that on the timescales of interest, there is no net production or degradation of NF-κB. It simply cycles in and out of the nucleus, i.e., the sum of nuclear and cytoplasmic NF-κB concentrations is a constant. Amongst the above listed processes, the association and dissociation of the complex $\{$NI$\}$ occurs fast enough that the concentration of the complex can be taken to be always in equilibrium with the free NF-κB and IκB concentrations. This allows us to describe the system using a very simple model consisting of only three variables [26], nuclear NF-κB (N_n), cytoplasmic IκB (I) and IκB mRNA (I_m):

$$\frac{dN_n}{dt} = A\frac{(1-N_n)}{\varepsilon+I} - B\frac{IN_n}{\delta+N_n}, \tag{7}$$

$$\frac{dI_m}{dt} = N_n^2 - I_m, \tag{8}$$

$$\frac{dI}{dt} = I_m - C\frac{(1-N_n)I}{\varepsilon+I}. \tag{9}$$

The external signal is supplied by IKK that enters the equations through the parameter, C, which is proportional to IKK concentration. The first term in Eq. (7) represents the import of free cytoplasmic NF-κB (whose concentration is $1-N_n$) into the nucleus. This is hindered by the presence of cytoplasmic IκB which sequesters NF-κB in the cytoplasm. Parameter A is proportional to the NF-κB nuclear

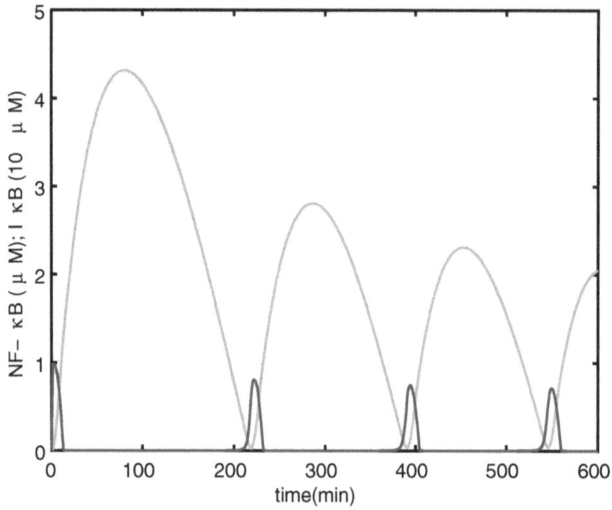

Fig. 7 Oscillations of nuclear NF-κB (N_n), red, and cytoplasmic IκB, green, for $A = 0.007$, $B = 954.5$, $C = 0.035$, $\delta = 0.029$ and $\varepsilon = 2 \times 10^{-5}$ (using parameter values taken from [3], see [26].) In order to facilitate comparison with the experimental plot, the x-axis has been limited to 600 min, but the oscillations are in fact sustained

import rate. The second term in Eq. (7) derives from the export of NF-κB from the nucleus via the NF-κB–IκB complex, which is why the term also depends on I. B is proportional to IκB nuclear import rate. δ sets the concentration at which half the nuclear IκB is complexed to NF-κB and it depends both on the rates of association and dissociation of the complex as well as the export rate. The first term in Eq. (8), the rate of production of IκB mRNA, contains the square of N_n because the production is activated by NF-κB dimers. The second term is the degradation of the mRNA whose rate sets the overall timescale. It is easy to modify this equation to deal with the β and ε isoforms of IκB simply by adding a constant for their NF-κB-independent rate of production. Equation (9) has one term for the production of cytoplasmic IκB from its mRNA and a second for its degradation due to the presence of IKK. This degradation is proportional to the concentration of the NF-κB – IκB complex, which depends on both I and $(1 - N_n)$, the concentration of cytoplasmic NF-κB. ε sets the concentration at which half of the cytoplasmic NF-κB is in the complex. C is proportional to the rate of degradation and to the IKK concentration.

Fig. 7 shows a plot of nuclear NF-κB and cytoplasmic IκB concentrations obtained in simulations, using parameter values from [3]. Our model predicts the following experimentally observed facts [3, 25]: (i) sustained oscillations in cells with only the α isoform of IκB, (ii) damped oscillations in wild type cells which include other isoforms of IκB, (iii) time period of the order of hours, (iv) spikiness of nuclear NF-κB and asymmetry of cytoplasmic IκB oscillations, (v) phase difference between NF-κB and IκB, (vi) lower frequency upon increased transcription of IκB.

Note that this is certainly not the only model able to reproduce the experimental oscillations of NFκB. Hoffman et al. have constructed a long list of chemical reactions between 26 different molecules in the NF-κB system, including 65 numerical parameters [3]. Krishna et al. have built an efficient model with seven variables and 11 numerical parameters [26]. Following the line described in the Introduction, we will discuss only the three–variable model described above, because only this is able to provide an insight into the physical mechanism which produce oscillations and guarantee robustness with respect to the numerical parameters.

2.5 Saturated Degradation of IκB

A key element in the model is the saturated degradation of cytoplasmic IκB in the presence of IKK (second term in Eq. (9)) due to the complex formation between NF-κB and IκB – a complex needed for IKK triggered degradation of IκB. The same complex inhibits nuclear NF-κB production because only free cytoplasmic NF-κB is imported into the nucleus.

A stability analysis of the system shows the importance of the saturated degradation for oscillations.

For small values of ε, corresponding to strong saturation of the degradation, the (unique) fixed point is unstable and the system goes into a periodic cycle. As ε is increased the fixed point becomes stable and the oscillations disappear (Fig. 8). This happens when the value of ε becomes comparable to the steady state value of I, which is precisely when the degradation rate stops being saturated. The saturation is crucial for oscillations because it puts an upper limit to the degradation rate, allowing IκB to accumulate and stay around longer than with the more usual I-proportional degradation rate. This effectively introduces a time delay into the feedback loop.

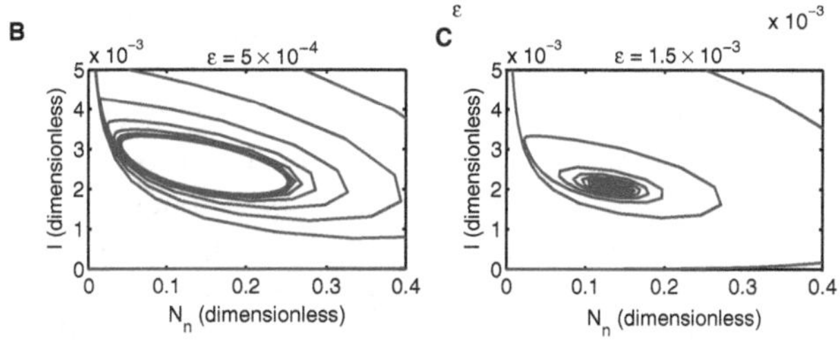

Fig. 8 (A) The stationary value of I as a function of the parameter ε which controls the binding between NFκB and IKK in the cytoplasm. (B) and (C) The projection of two trajectories at different values of ε onto the $N_n - I$ plane

Interestingly, the NF-κB core in Fig. 2c is similar to an early model showing oscillations by negative feedback [27] which introduced precisely the same kind of saturated degradation to mitigate the unreasonably large Hill coefficient in an even earlier model of Goodwin [28]. The importance of this mechanism has also been recognized by Goldbeter who has used it in models of various cellular oscillations, e.g., the cell cycle [29], development in myxobacteria [30], yeast stress response [31] and the mammalian circadian clock [32]. Saturated degradation has also been implicated in models of calcium oscillations in cells [33, 34], suggesting that saturated degradation might be a very general mechanism, easily implemented by complex formation and used by cellular processes to introduce time delays where necessary.

2.6 Properties of the NF-κB Oscillation

What functional role, if any, do the oscillations in the NF-κB system play? There have been several suggestions: downstream gene networks are perhaps regulated by the frequency of the oscillations, or the oscillations could be a by-product of rapid attenuation of NF-κB, or they might be used to make multiple evaluations of the input signal [35, 36]. Barken et al. [37] warn against overemphasizing the physiological role of oscillations. We will show that the spiky oscillations can, indeed, show a high sensitivity to IKK. This sensitivity allows a great versatility in the regulation of downstream genes by NF-κB. Where the cell requires a gene to be very sensitive to the IKK concentration, the NF-κB system can result in steep response curves with Hill coefficients larger than 20. And where a slower response is necessary, it can be achieved by adjusting the binding and dissociation constants of NF-κB to that operator site. Further, we will show that cascades of different length could be used to turn certain genes on earlier or later. Given this versatility in regulatory strategies it seems likely that cells would have evolved to make use of these properties of the NF-κB oscillations. It remains for future experiments to uncover the particular ways NF-κB regulates specific genes.

Since IKK is the external signal to which the system responds, we begin by comparing the sensitivity of spiky and soft oscillations to changes in IKK concentration. We consider two quantities: the spike duration, defined as the amount of time N_n spends above its average value, and the spike peak, defined as the maximum nuclear NF-κB concentration during each cycle of oscillations. Figure 9A shows how the spike duration depends on IKK concentration. The sensitivity of the spike duration is very high in certain regions of spiky oscillations. It is especially large near the transition to soft oscillations. A similar sensitivity is seen in the peak NF-κB concentration (Fig. 9B). Thus, the spike duration and peak are much more sensitive to (and therefore easier to regulate by) IKK for spiky than for soft oscillations.

It is possible for genes regulated by NF-κB to inherit this sensitivity in the form of a high effective Hill coefficient.

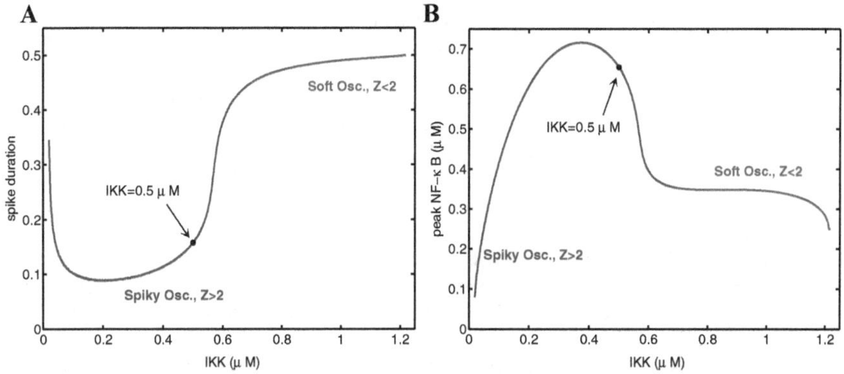

Fig. 9 Sensitivity to IKK. **A** Spike duration, the fraction of time N_n spends above its mean value, as a function of IKK concentration. The black dot shows the IKK value used in Fig. 7. Blue and red signify, respectively, regions of spiky and soft oscillations. Notice the sharp response just before the transition to soft oscillations. **B** Spike peak, the maximum concentration of nuclear NF-κB, as a function of IKK concentration. The black dot shows the IKK value used in Fig. 7. Blue and red signify, respectively, regions of spiky and soft oscillations. Notice the sharp response just before the transition to soft oscillations

3 Summary and Outlook

The oscillating character of gene expression in a number of systems challenges the traditional framework, where genetic control is operated under equilibrium conditions. This fact makes the modelling of these biological systems more difficult, and thus the tools needed to investigate their kinetics has to be more careful. In spite of these difficulties, one can look for basic physical principles which are at the basis of expression oscillations. These seem to be the presence of feed–back loops, of delayed interaction and of saturated binding.

An obscure point concerning expression oscillations is what is their purpose. A trivial answer could be that they are an unavoidable byproduct of small feed–back loops. This seems unlikely to be the correct answer, because oscillations are able to carry much more information than mere equilibrium–concentration shifts, and it would be very anti–economical for evolution to waste such a rich potential.

References

1. Haupt Y, Maya R, Kazaz A and Oren M (1997) Mdm2 promotes the rapid degradation of p53, *Nature* **387**, 296–299.
2. Hirata H, Yoshiura S, Ohtsuka T, Bessho Y, Harada T, Yoshikawa K and Kageyama R (2002) Oscillatory expression of the bHLH factor Hes1 regulated by a negative feedback loop, *Science* **298**, 840–843.

3. Hoffmann A, Levchenko A, Scott M L and Baltimore D (2002) The IkappaB-NF-kappaB signaling module: temporal control and selective gene activation, *Science* **298**, 1241–1245.
4. Barkai N and Leibler S (2000) Circadian clocks limited by noise, *Science* **403**, 267–268.
5. Yeger–Lotem E, Sattath S, Kashtan N, Itzkovitz S, Milo R, Pinter R Y, Alon U and Margalit H (2004) Network motifs in integrated cellular networks of transcriptionregulation and proteinprotein interaction, *Proc. Natl. Acad. Sci. USA* **101**, 5934–5939.
6. Strogatz S (1994) *Nonlinear Dynamics and Chaos* (Addison-Wesley, Reading MA).
7. Thomas R (1981) *Quantum noise*, Springer Series in Synergetics 9, Ed. Gardiner, Springer, Berlin, pp. 180–193.
8. Gouzé J L (1998) Positive and negative circuits in dynamical systems, *J. Biol. Syst.* **6**, 11–15.
9. Bar–Or R L, Maya R, Segel L A, Levine A J and Oren M (2000) Generation of oscillations by the p53–Mdm2 feedback loop: a theoretical and experimental study, *Proc. Natl. Acad. Sci. USA* **97**, 11250–11256.
10. Schon O, Friedler A, Bycroft M, Freund S M V and Fersht A R (2002) Molecular mechanism of the interaction between MDM2 and p53, *J. Mol. Biol.* **323**, 491.
11. Mallet-Paret J and Smith H L (1990) The Poincare-Bendixson theorem for monotone cyclic feedback systems, *J. Dyn. Diff. Eq.* **2**, 367–421.
12. Mallet-Paret J and Sell G R (1996) Systems of delay differential equations I: Floquet multipliers and discrete Lyapunov functions, *J. Differ. Eq.* **125**, 441–489.
13. Goldbeter A et al. (2001) From simple to complex oscillatory behavior in metabolic and genetic control networks, *Chaos* **11**, 247–260.
14. Picksley S M and Lane D P (1993) The p53–Mdm2 autoregulatory feedback loop: a paradigm for the regulation of growth control by p53? *Bioessays* **15**, 689–690.
15. Lahav G, Rosenfeld N, Sigal A, Geva–Zatorosky N, Levine A J, Elowitz M B and Alon U (2004) Dynamics of the p53–Mdm2 feedback loop in individual cells, *Nature Genetics* **36**, 147.
16. Gottlieb T and Oren M (1996) p53 in growth control and neoplasia, *Biochim. Biophys. Acta* **1287**, 77.
17. Tiana G, Jensen M H and Sneppen K (2002) Time delay as a key to apoptosis induction in the p53 network, *Eur. Phys. J. B* **29**, 135–139.
18. Neamtu M, Horhat R F and Opris D (2006) A dynamic p53-mdm2 model with delay kernel, *arXiv preprint* arXiv.math.DS/0601481.
19. Alberts B, Bray D, Lewis J, Raff M, Roberts K and Watson J (1994) *Molecular Biology of the Cell.* Garland: Science, Taylor&Francis, New York.
20. Ribbeck K and Gorlich D (2001) Kinetic analysis of translocation through nuclear pore complexes, *EMBO J.* **20**, 1320–1330.
21. Greenblatt M S, Bennett W P, Hollstein M and Harris C C (1994) Mutations in the p53 tumor suppressor gene: clues to cancer etiology and molecular pathogenesis, *Cancer res.* **54**, 4855–4878.
22. Kageyama R, Ishibashi M, Takebayashi K and Tomita K (1997) bHLH transcription factors and mammalian deuronal differentiation, *Int. J. Biochem. Cell Biol.* **29**, 1389.
23. Jensen M H, Sneppen K and Tiana G (2003) Sustained oscillations and time delays in gene expression of protein Hes1, *FEBS Lett.* **541**, 176–177.
24. Pahl H L (1999) Activators and target genes of Rel/NF-kappaB transcription factors, *Oncogene* **18**, 6853–6866.
25. Nelson D E, Ihekwaba A E C, Elliott M, Johnson J R, Gibney C A, Foreman B E, Nelson G, See V, Horton C A, Spiller D G et al. (2004) Oscillations in NF-B signaling control the dynamics of gene expression, *Science* **306**, 704.
26. Krishna S, Jensen M H and Sneppen K (2006) Minimal model of spiky oscillations in NF-kB signaling, *Proc. Natl. Acad. Sci. USA* **103**, 10840–10845.
27. Bliss R D, Painter P R and Marr A G (1982) Role of feedback inhibition in stabilizing the classical operon, *J. Theor. Biol.* **97**, 177–193.
28. Goodwin B C (1965) *Adv. Enzyme Regulation*, ed. Weber, G. (Pergamon Press, Oxford) Vol. 3, pp. 425–438.

29. Goldbeter A (1991) A minimal cascade model for the mitotic oscillator involving cyclin and cdc2 kinase, *Proc. Natl. Acad. Sci. (USA)* **88**, 9107–9111.
30. Igoshin O A, Goldbeter A, Kaiser D and Oster G (2004) A biochemical oscillator explains several aspects of Myxococcus xanthus behavior during development, *Proc. Natl. Acad. Sci. USA* **101**, 15760–15765.
31. Jacquet H, Renault G, Lallet S, Mey J D and Goldbeter A (2003) Oscillatory nucleocytoplasmic shuttling of the general stress response transcriptional activators Msn2 and Msn4 in Saccharomyces cerevisiae, *J. Cell Biol.* **161**, 497–505.
32. Leloup J C and Goldbeter A (2003) Toward a detailed computational model for the mammalian circadian clock, *Proc. Natl. Acad. Sci. USA* **100**, 7051–7056.
33. Reidl J, Borowski P, Sensse A, Starke J, Zapotocky M and Eiswirth M (2005) Model of calcium oscillations due to negative feedback in olfactory cilia, *Biophys. J.* **90**, 1147–1155.
34. Goldbeter A, Dupont G and Berridge M (1990) Minimal model for signal-induced Ca2+ oscillations and for their frequency encoding through protein phosphorylation, *Proc. Natl. Acad. Sci. USA* **87**, 1461–1465.
35. Lahav G (2004) The strength of indecisiveness: oscillatory behavior for better cell fate determination, *Science's STKE* pe55.
36. Ting A Y and Endy D (2002) Decoding NF-kB signaling, *Science* **298**, 1189–1190.
37. Barken D, Wang C J, Kearns J, Cheong R, Hoffmann A and Levchenko A (2005) Comment on "Oscillations in NF-kB Signaling Control the Dynamics of Gene Expression", *Science* **308**, 52a–52a.

Translation Attenuation Mechanism in Unfolded Protein Response

Ala Trusina, Feroz Papa and Chao Tang

Abstract Endoplasmic Reticulum is a cellular organelle where membrane and extracellular proteins are folded with the help of chaperons. Insulin is one example of such extracellular proteins. Unfolded Protein Response (UPR) is a cell response to an increased level of unfolded proteins in ER. In pancreatic β-cells failure in UPR leads to accumulation of unfolded insulin in Endoplasmic reticulum and eventual cell death. This is thought to be one of the causes of type two diabetes.

Keywords UPR, stress response, translational attenuation, unfolded protein response

1 Two Strategies to Deal with the ER Stress:

1.1 Upregulate Chaperons Through Ire1 Pathway

One of the conserved strategies of UPR is to upregulate chaperons and proteases to deal with the increased levels of unfolded proteins. This mechanism is well conserved from yeast to mammals and is activated through Ire1 pathway (see Fig. 1). Ire1 protein works as a sensor of an increased concentration of Unfolded Proteins either due to misfolding of already folded proteins or influx of newly translated peptides. Ire1 is activated through dimerization and subsequent autophosphorylation. The dimerization is facilitated by Unfolded proteins and is obstructed by chaperons, which can bind to a dimerization interface. Once activated, Ire1 can work as endonuclease and splice a specific substrate – messenger RNA of a Hac1 Transcription Factor (TF). Interestingly, in yeast, unspliced Hac1 mRNA cannot be translated even though it is found to have ribosomes associated to it. The reason is that the unspliced mRNA can form a hairpin loop, and those ribosomes that initiated translation

Ala Trusina, Feroz Papa and Chao Tang
QB3, University of California at San Francisco

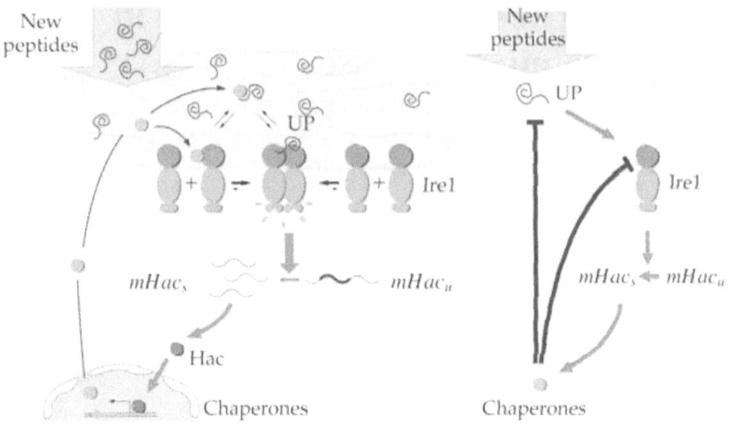

Fig. 1 Ire1 pathway. Increase in Unfolded proteins activates Ire1 either directly by facilitating dimerization of its ER domains and further autophosphorylation, or indirectly by sequestering the chaperons that can bind to Ire ER domains and prevent dimerization. Chaperons are drown as blue circles and red spirals stand for the unfolded proteins. Wavy lines show Hac mRNA, red insert shows the intron in unspliced Hac1 mRNA. Right panel summarizes the process with green lines showing positive upregulation, and red lines negative regulation

before the loop has formed are being trapped and can only continue translation once the intron is spliced out by Ire1. This can explain almost immediate increase in Hac1 protein upon ER stress (75% of splicing occurs within first 15 min after stress), the pool of "ready-to-go" inactive mRNA is converted into pool of active mRNA with ribosomes already pre-assembled. Once translated transcription factor Hac1 is translocated into the nucleus and upregulates transcription of chaperons and proteases to decrease the concentration of unfolded proteins. Interestingly that even though the Transcription Factor corresponding to Hac1 in mammalian cells does not share any sequence similarity, it is upregulated in very much similar way as Hac1, through its mRNA splicing by Ire1. The rest of the components in Ire1 pathway are conserved from yeast to mammals. There is another pathway in mammalian cells that upregulates chaperons (ATF6 pathway), but we will not discuss it here for the purpose of clarity.

1.2 Decrease Flux of New Peptides into the ER through Transaction Attenuation Pathway

Through evolution multicellular organisms have evolved an additional strategy to deal with stresses in ER. A pathway evolved that stops the translation of new proteins into the ER in response to increased amount of Unfolded Proteins. PERK (Pancreatic Endoplasmic Reticulum Kinase) plays a role of the stress sensor in this pathway. Interestingly, its ER domain is homologous to Ire1 domain and it is believed the sensing mechanism is similar to that of Ire1. As in Ire1, dimerized PERK

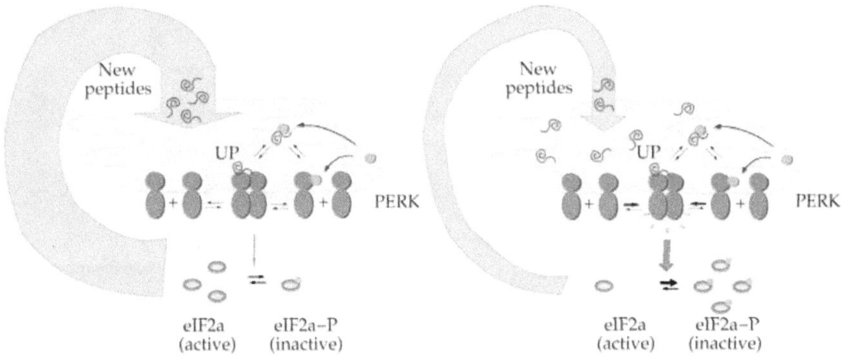

Fig. 2 Translation Attenuation (TA) pathway. Increase in Unfolded proteins activates PERK in similar way as Ire1. Left panel shows the pathway in normal conditions: without ER stress the flux of the new peptides is large, eIF2a is mostly in unphosphorylated form. Middle panel shows the response to stress: Perk is activated, eIF2a is phosphorylated, the flux of new peptides is reduced

is autophosohorylated, however cytoplasmic domain of Perk is distinct form that of Ire1 and serves totally different purpose. In eukaryotes for translation to initiate, ribosomes should form a ternary complex with eIF2α protein. Such a complex can only be formed with unphopsphorylated eIF2α. Thus phopsphorylation of eIF2α downregulates overall translation and it is exactly this is what activated PERK does: it phosphorylates eIF2a in response to stress and thus downregulates the influx of new peptides into the ER (see Fig. 2). It is important to note that Translation Attenuation pathway is activated first, within first 30 min after stress, whereas Ire1 pathway is activated last, after about 6 h after stress.

2 Is There Translation Attenuation Pathway in Yeast?

2.1 Mathematical Modeling

It is not clear if Translation Attenuation in response to ER stress takes place in yeast. There are contradictory reports in literature, in one case the phosphorylation of eIF2a was observed when yeast was treated with extreme doses of Tunicamycin, a chemical agent commonly used to cause ER stress (Cherkasova et al. [2]) in another report the phosphorylation of eIF2a remained constant, however the levels of the GCN4 protein which is upregulated when eIF2a is phosphorylated went up three fold [4]. This upregulation of GCN4, seems to be Ire1 pathway dependent, since in Ire1 deletion strain no upregulation was observed. One possible explanation of Ire1 dependence is formation of stable GCN4:Hac1 complex (both GCN4 and Hac proteins are short lived, with half lives of about five minutes, see Fig. 3).

This dependence on Ire1 pathway excludes possibility that GCN4 is upregulated through Translation attenuation pathway, however the GCN4 upregulation could have been missed in Patil et al. experiment if in Ire1 deletion strain GCN4 peaks earlier than the first time point at 15 min.

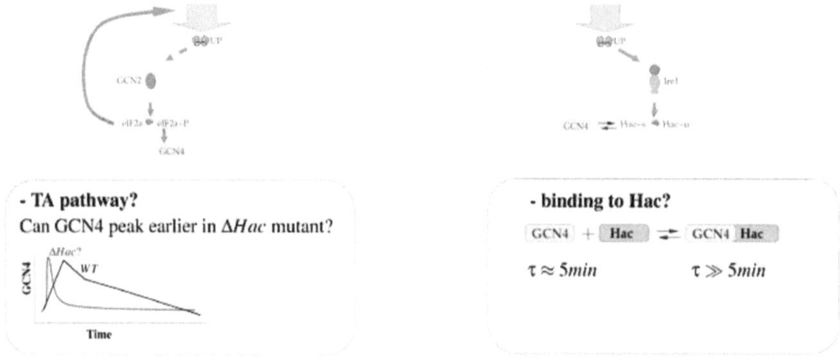

Fig. 3 Addressing the hypothesis based on Patil et al. [4] data through mathematical modeling. The questions we are addressing are: Is GCN4 transient upregulation is due to Translation Attenuation pathway (left) or through binding to Hac and stabilization (right)?

To address this issue we have built mathematical model and asked two questions: Firstly, If GCN4 is upregulated through Translation attenuation pathway is it theoretically possible to get GCN4 peaking earlier in Ire1 deletion strain compared to wt? Secondly, if GCN4 transient induction is due to stabilization through Hac1, how can GCN4 go down to the steady state level it had before stress whereas Hac1's steady state level is higher than that before the stress? These questions are summarized in Fig. 4.

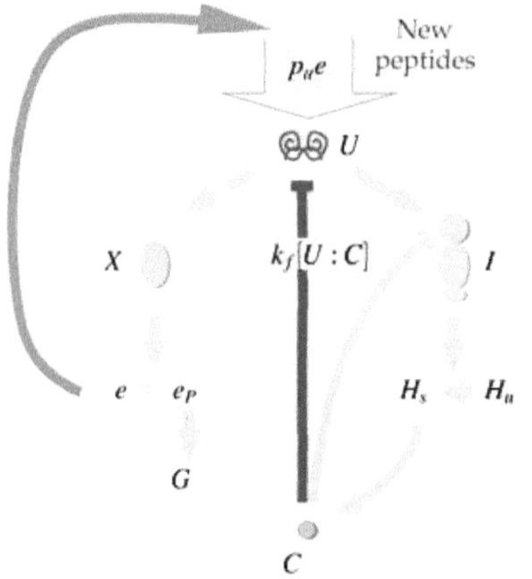

$$\frac{dU}{dt} = p_u e + k_u F - k_f[U:C] - \frac{U}{\tau}$$
$$\frac{dF}{dt} = -k_u F + k_f[U:C] - \frac{F}{\tau}$$

$$\frac{dH_u}{dt} = p_H - b[I:U] - \frac{H_u}{\tau_1}$$
$$\frac{dH_s}{dt} = b[I:U] - \frac{H_s}{\tau_2}$$

$$\frac{dG}{dt} = p_G + a_g e_p - \frac{G}{\tau_G}$$

$$\frac{dC}{dt} = p_C \frac{H_s}{H_s + K_H} - \frac{C}{\tau}$$

Fig. 4 Addressing the hypothesis based on Patil et al. [4] data through mathematical modeling. Left panel illustrates the network use used to built mathematical model. Each equation describes how the concentration of each of the components/nodes changes in time, for example the rate of the increase in unfolded proteins, U, is determined by the translation rate, $p_U e$ and we assume it depends linearly on concentration of eIF2a, e. This term corresponds to the green link from eIF2a to Unfolded Proteins in left panel. Unfolding of the folded proteins, F, positively contribute with $k_u F$, to the Ingres in U. The rate of decrease in U is determined by the chaperon, C, folding term $k_f[U:C]$ (red link from C to U) and dilution through division U/τ, where τ is yeast doubling time

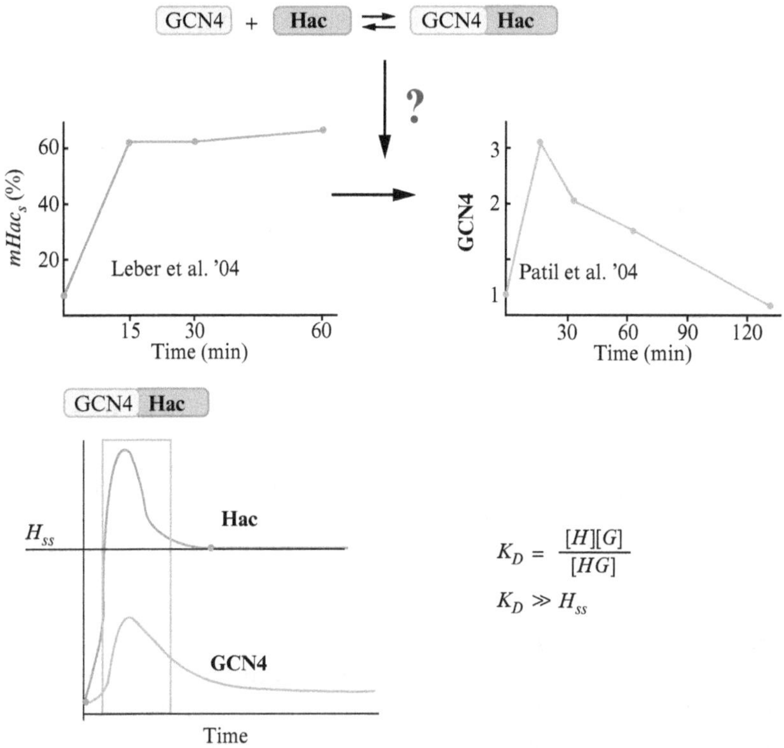

Fig. 5 Is GCN4 transient upregulation due to stabilization through Hac1 binding? If yes, then how can levels of GCN4 increase transiently (orange curve, right) whereas Hac1 mRNA reaches new steady state levels [3] (green curve, left)? The possible solution is "transient stabilization" that requires significant overshoot in Hac1 protein levels. An additional condition is that the binding between two proteins is week enough to allow GCN4 to be in its free form when at steady state

The model is summarized in Fig. 4 and the details will be published elsewhere. Through our modeling approach we came to a conclusion that

(a) The mathematical model based on the network shown in Fig. 4 cannot generate peak in GCN4 that will be earlier in Ire1/Hac1 deletion strain than in wild type strain. This is primarily due to the fact that the rate of the increase in GCN4 will be determined by the level of stress and it indeed can be higher in Ire1 deletion strain, however the rate of decrease in GCN4 is determined by how fast is the recovery from stress happens and will always be faster when both Ire1 and TA pathways work together than when TA works alone in Ire1 deletion strain, $\Delta Ire1$. Thus the rate of decrease in GCN4 will always be smaller in $\Delta Ire1$ and GCN4 will peak later than in wild type strain.
(b) The transient increase in GCN4 can only be the result of stabilization through Hac1 if Hac1 kinetics has a prominent overshoot and at the same time binding between GCN4 and Hac1 is week enough so that GCN4 does not bind to Hac1 in absence of stress and also almost all of GCN4 is in its free from when Hac1 reaches its new steady state level. This is summarized in Fig. 6.

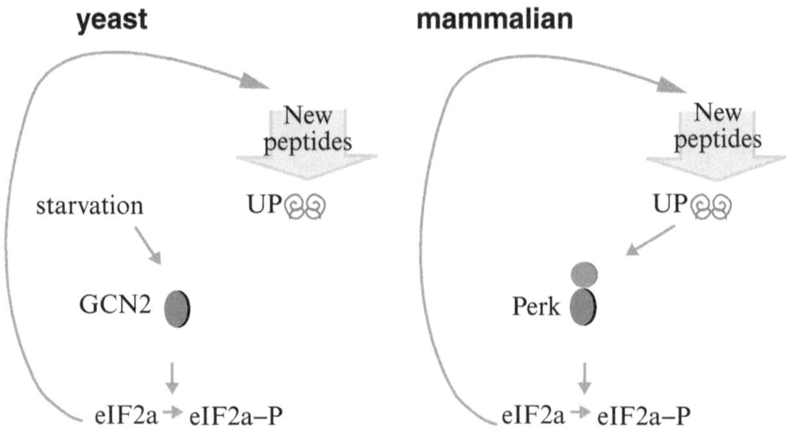

Fig. 6 It is not clear if Translation Attenuation in response to ER stress takes place in yeast. It is certain that PERK protein is not present, however yeast has GCN2 kinase that is homologous to PERK kinase domain. In yeast Translation Attenuation is used in response to aminoacis starvation stress

The second assumption is indirectly supported by the particular activation mechanism of Hac through Ire1 splicing. As we mentioned earlier, there is a pool of "inactive" mRNA with trapped ribosomes and it is very plausible that it can cause such an overshoot in protein levels of Hac1 taken into account its very short half life estimated to be 1.5–3 minutes. In addition, the work by Bock-Axelsen et al. [1], investigates the overshoot in Hac1 protein levels and suggests it to be an efficient way to speed up the upregulation of chaperons.

It is certain that PERK protein is not present, however yeast has GCN2 kinase that is homologous to PERK kinase domain. Through GCN2 Translation Attenuation is employed in response to aminoacis starvation stress (see Fig. 5). Thus PERK is a cut-and-paste product of evolution. It seemingly was combined out of two proteins present in Yeast, Ire1 lumenal domain and GCN2 kinase domain and this combination links the Translation Attunation startegy with the ER stress.

To address the question of why didn't yeast evolve a protein similar to PERK and wether there yeast will benefit from acquring it, we have syntheticaly engineered yeast analog of PERK protein by combining yeast IRE1 ER domain with mouse PERK kinase domain as shown in Fig. 7. Ire1 is necessary for yeast to survive ER stresses. Thus strain where this protein and is deleted and thus the whole Ire1 pathway is disabled does not grow when subjected to ER stress. (In our experiments we used agar plates with Tunicamycin spread on agar surface at concentration of 0.12 ug/ml. Tunicamycin prevents proper protein folding in ER.)

Interestingly, when chimera protein, yPERK, was introduced in yeast lacking Ire1, yeast cells could survive stress, see Fig. 8. Yeast cells with Ire1 deleted from the chromosome grow only if we insert Ire1 back (Ire1 on a high copy plasmid, second from left) or if we insert yPerk (right panel). We didn't detect any obvious advantage of yPERK's presence in wild type strain. This interesting result confirms

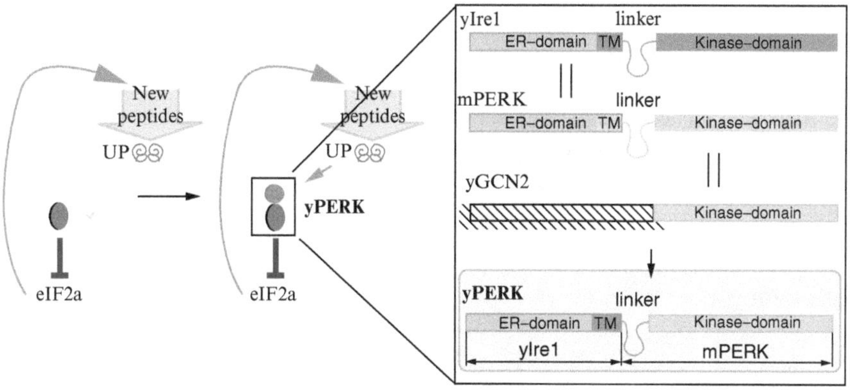

Fig. 7 Engineering synthetic TA pathway in yeast by cutting-and-pasting from the existing components. Yeast IRE1 Endoplasmic Reticulum domain was fused to mouse PERK kinase domain

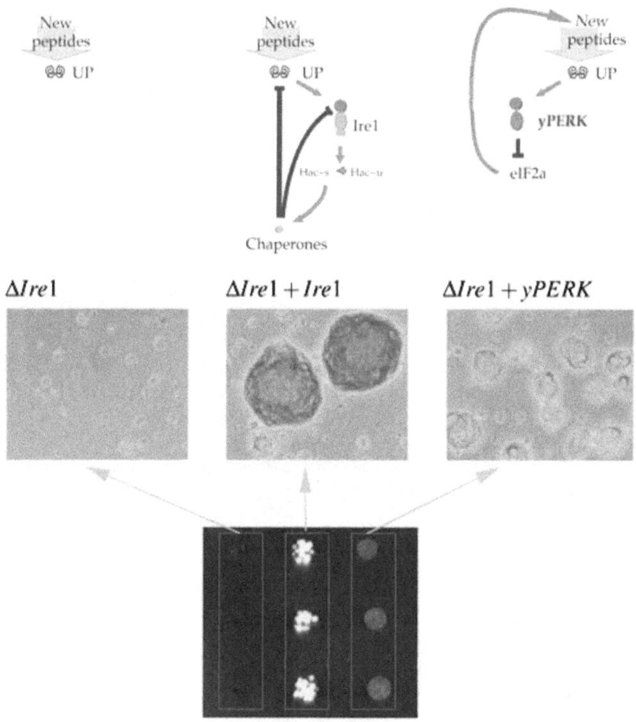

Fig. 8 yPERK chimera protein rescues Ire1 deletion strain. Top panel illustrates which pathways are present in each of the strains in the middle panel. Middle panel shows the microscopy pictures of the yeast colonies in Ire1 deletion strain, Ire1 deletion strain complemented with Ire1 protein on high copy plasmid, and Ire1 deletion strain complemented with yPerk chimera protein on high copy plasmid. Bottom panel is the picture of the plate, each white circle being a spot of growing yeast colonies

that yeast can use TA strategy to deal with ER stress and opens for a question of why didn't it evolve this mechanism that seems to be so necessary in mammalian secretory cells? Did yeast avoid evolving PERK? What differences in environment and function can account for the fact that yeast does not use PERK in UPR? These and other questions remain to be answered either through experimental approach or, whenever possible, through modeling.

References

1. Axelsen, J. B. and Sneppen, K. (2004) Quantifying the benefits of translation regulation in the unfolded protein response, *Phys. Biol.* **1**, 1–7.
2. Cherkasova, V. A. and Hinnebusch, A. G. (2003) Translational control by TOR and TAP42 through dephosphorylation of eIF2alpha kinase GCN2, *Phys. Biol.* **17**, 859–872.
3. Jess H. Leber, Sebastian Bernales, P. W. (2004) IRE1-independent gain control of the unfolded protein response, *Plos Biol.* **2**, 235.
4. Patil (2004) Gcn4p and novel upstream activating sequences regulate targets of the unfolded protein response, *Plos Biol.* **2**, 246.

The Origin and Evolution of Viruses

Vadim I. Agol

Abstract The lecture covers three main topics: (i) Viruses: properties, place in the living world, and possible origin; (ii) Molecular basis of viral variability and evolution; and (iii) Evolution of viral pathogenicity and emerging viral infections.

Keywords Selfish genetic elements, replication, transcription, mutation, recombination, plasmids, pathogenicity, fibroma/myxoma virus, influenza virus, SARS-coronavirus, human immunodeficiency virus

1 Viruses: Properties, Place in the Living World, and Possible Origin

Viruses are selfish genetic elements existing in two forms, the passive extracellular form represented by a molecule(s) of nucleic acid (DNA or RNA) usually enclosed into a protein-containing shell, and the active intracellular replicating and gene-expressing form. Viruses are the most abundant biological entities on the Earth. According to some estimates, our planet harbors 10^{31} virus particles (Breitbart and Rohwer, 2005). There are $\sim 2 \times 10^8$ ton carbon in marine viruses only (=carbon content in 7×10^7 blue whales). If stretched end to end, marine viruses would span $\sim 10^7$ light years ($\sim$100 times the distance across our galaxy) (Suttle, 2005). The shape of virus particles is highly variable (e.g., rod-like, thread-like, with cubic symmetry, and many others). The viral genomes may be represented by either single-stranded or double-stranded DNA or RNA molecules, which may have either linear

Vadim I. Agol
M. P. Chumakov Institute of Poliomyelitis & Viral Encephalitides
Russian Academy of Medical Sciences and
A. N. Belozersky Institute of Physical-Chemical Biology
M. V. Lomonosov Moscow State University
Moscow Russia

or circular structures. The number of viral genes may vary from one (e.g., in hepatitis delta virus) to ~1,200 (in mimivirus). In the latter case, the size of the viral genome is several-fold larger than that in the smallest cellular organisms.

A distinctive property of viruses is the absence of protein-synthesizing and energy-generating machineries. As a result, viruses are strictly dependent on the host cell. Another fundamental distinction between viruses and cells is the mode of their multiplication. Viral reproduction is based on disjunctive mechanisms, whereupon virus-specific proteins and genomic nucleic acids are accumulated as separate pools, from which multiple mature progeny viral particles (or their core nucleoproteins) are eventually assembled. On the other hand, cellular multiplication involves (usually binary) division. During a part of their life cycle, the genome of some viruses may be integrated into the host cell genome, existing and dividing as a cellular chromosomal segment.

Depending on the type of genome and mechanisms of its replication and expression, several major types of viral "strategies" can be discerned (Baltimore, 1971; Agol, 1974). Remarkably, all the theoretically possible replication/transcription systems based on the principle of complementarity appear to be exploited by viruses. The replication/transcription system used by genomes of cellular organisms comprises only a minor subset of the systems exploited by viruses.

Viruses have very close relatives, selfish genetic elements (DNA or RNA): viroids (small infectious non-coding circular RNA), plasmids (various non-chromosomal DNA or RNA elements), transposons (DNA elements moving between different positions of a genome; mobile genetic elements), retrotransposons (mobile DNA elements exploiting reverse transcription) and some others. Viruses alike, they are fully dependent on cellular translation and energy-generating machineries and use the same expression and replication mechanisms as viruses do. They may be integrated into cellular DNA or RNA or exist separately, may be amplified in a cell, move between different parts of the genetic material of a cells or migrate between cells. Therefore, these elements can be combined with viruses into a common domain of life. In distinction from viruses, however, they lack protein coats and stable extracellular forms.

There are two major problems related to the origin of viruses. What are relationships between viruses and cellular organisms? In other words, is there a place(s) for viruses on universal trees of life proposed for its three kingdoms, Archea, Eubacteria, and Eukarya? And secondly, are viruses monophyletic, i.e., is there a single evolutionary tree or branch of viruses? During the first century after discovery of viruses, several purely speculative but quite imaginative hypotheses had been proposed: according to them, viruses originated from escaped ("mad") cellular genes, or from degenerated cells, or they are descendants of precellular genetic elements.

Modern hypotheses of viral origin are based on two major developments of the molecular biology: discovery of ribozymes (RNA-based enzymes) and formulation of the "RNA World" theory (RNA had been "invented" before proteins and DNA), on the one hand, and achievements of genomics (determination of the nucleotide sequences of a great number of cellular and viral genomes), on the other. These hypotheses postulate a very significant contribution of viruses to the genetic information of cellular organisms.

According to the "Three viruses, three domains" hypothesis proposed by P. Forterre, the cellular Last Universal Common Ancestor (LUCA) had an RNA genome and harbored a variety of already created RNA-containing viruses. During the next step, DNA and DNA viruses had been invented. Three distinct DNA viruses, which had infected RNA genome-containing cells, gave rise to the three distinct domains of life, bacteria, archea, and eukarya (Forterre, 2006). A detailed hypothetical evolutionary scenario was described by E. Koonin and colleagues (Koonin and Martin, 2005; Koonin et al., 2006). The authors suggest that life has originated in inorganic compartments serving as surrogate cells. RNA molecules with enzymatic activities (ribozymes) could attain there relatively high concentrations and could move between compartments and "infect" them. At this precellular stage, several fundamental inventions had been consecutively made. The invention of proteins led to the appearance of viruses with positive single-strand and with double-strand RNA genomes. The invention of DNA resulting from RNA-dependent DNA synthesis led to the appearance of retroviruses, retroid viruses and retroposons. Finally, DNA-dependent DNA synthesis had been invented, giving rise to the appearance of viruses with double-strand DNA genomes and DNA plasmids. Only after achieving such genetic variability and richness, the "pre-archeal" and "pre-bacterial" inorganic compartments had been replaced by cellular forms, which retained numerous ancient viruses. The engulfment of a bacterium by an archaeon was the starting point of eukarya, which acquired viruses of both of their "parents". According to Koonin, the major classes of viruses do not have a common origin in the traditional sense but neither are they unrelated. On the other hand, a very significant proportion of the cellular genetic material has evolutionary relatedness to the genetic composition of the ancient world of RNA viruses. It is estimated that well over a half of mammalian genome is inherited from viruses. In other words, "the tree of life and its root are immersed in a viral ocean" (Bamford, 2003).

The above scenarios also suggest that modern viruses have inherited molecular mechanisms that have disappeared from modern DNA cells. That is why transcription and replication mechanisms in the viral world are more diverse than those in the cellular world. It cannot be excluded that many yet unknown molecular mechanisms exist in the current viral world. The exploration of the viral diversity is one of the major challenges of biology in this century.

2 Molecular Basis of Viral Variability and Evolution

After their "birth", viral genomes had been and are still evolving by accumulation of point mutations and genome rearrangements (duplications, deletions, recombination). The frequency of mutations in DNA viruses may be several orders of magnitude lower than in RNA viruses, whose replicative enzyme, the RNA-dependent RNA polymerase, lacks proof-reading activity. The error frequency in RNA viruses is also variable, and, in some of them, any newly synthesized genomic RNA molecule contains on an average a mutation. Such viruses exist on the edge of mutational catastrophe: even a few-fold increase in the error frequency may result in population extinction. Nevertheless, a certain level of replicative infidelity is highly advantageous for viruses since it confers a potential for changing and adaptation. For example, replicative errors are one of the major factors contributing to the development of drug-resistance and hence to the scarcity of efficacious antiviral drugs. Interestingly, a single point mutation in the poliovirus RNA-dependent RNA polymerase increases its fidelity several-fold. However, the more accurate mutant is less fit: not only it develops drug-resistance less readily, but is also less neurovirulent (Pfeiffer and Kirkegaard, 2003, 2005; Vignuzzi et al., 2006). A heterogeneous population, as a group, may have advantages over a more homogeneous population. The advantages may be due not only to a greater potential for adaptation but also to the possibility of mutual cooperation.

The genome instability challenges the very identity of viruses: how does virus X remain virus X? However, despite their intrinsic instability, viral genomes, even in the extreme case of RNA viruses, are remarkably robust. Several factors contribute to the genetic stability of viruses in nature. Adverse mutations are eliminated by negative selection, whereas fitness-increasing mutations are nearly not-existing in constant environment. Adaptive mutations may be positively selected upon environmental changes (including infection of a novel host species). The only major factor favoring accumulation of neutral mutations is bottlenecking (stochastic picking up a single viral particle, or a few of them, from a heterogeneous population), which is quite common during natural transmission of viruses. Consecutive bottlenecking events may result in a more or less marked decrease in viral fitness (the so called Muller's ratchet) because viral genomes picked up by chance from a heterogeneous population may well harbor detrimental mutations.

To what extent is the Muller's ratchet inevitable and irreversible? The robustness of viruses may be studied experimentally by damaging their genome (e.g., by introducing point mutations or deletions, randomization of portions of the nucleotide sequence, or even breaking the genome apart) followed by investigating viral viability and fitness as well as structure and stability of the mutated viral genome. A wealth of relevant information concerning numerous viruses has already been accumulated in literature. Some examples of such "interrogation under torture" carried out with picornaviruses in the author's laboratory are presented below.

The Theiler's murine encephalomyelitis virus (TMEV) is highly pathogenic for mice under certain conditions: a few wild type viral particles injected intracerebrally kill mice. For efficient viral reproduction in the brain, interaction of a translational control element (IRES) in the 5′-noncoding region (5NCR) of the viral RNA with a host protein, neural polypyrimidine tract-binding protein (nPTB), is essential (Pilipenko et al., 2001). A single point mutation in one of the nPTB-binding sites of the viral RNA led to a dramatic decrease in viral neurovirulence: $\sim 10^4$ more mutant particles were now needed to kill a mouse. However, wild-type level of neurovirulence could readily be restored by either reversion or pseudoreversions during viral reproduction, e.g., by generation of a new nPTB-binding site in the viral 5NCR due to a compensating point mutation (Pilipenko et al., 2001). An 8-nt deletion in the 5NCR of poliovirus RNA severely impaired viral fitness and nearly killed the virus. However, this defect could be readily fixed by natural pseudoreversions of at least three different kinds, including point mutations, insertions and extended deletions (Pilipenko et al., 1992; Gmyl et al., 1993).

The results of these and other experiments show that many mutations and their combinations are life-compatible and may not decrease viral fitness and, if they have adverse effects, these effects may often be repaired or compensated. Even broken molecules of genomic RNA may sometimes be repaired by recombination (Gmyl et al., 1999, 2003). Thus, viral RNA genomes are robust and exhibit remarkable vitality and phoenix-like phenotype. What does then force them to evolve and to form new species and genera?

A hypothesis is put forward according to which a significant decrease in fitness (caused for example by mutations or host changes) may result in genome instability, which in turn would produce a set of various low-fit variants. Under non-competitive conditions (e.g., in a small-sized population), such variants may serve as avid acceptors of novel elements acquired by recombination or by some other genetic modifications. Further fine genetic tuning may convert the new creature into a well fit new viral species. Thus, evolutionary "jumps" may not necessarily be due to a consecutive acquisition of improving mutations but rather they may result from injury of an important element followed by metastability of the viral genome, and acquisition of a novel element.

3 Evolution of Viral Pathogenicity and Emerging Viral Infections

Viruses are not only major human, animal, and plant pathogens, sometime deadly, being the causative agents of small pox, HIV, influenza, foot-and-mouth disease and many other severe pathological conditions, but also they are a driver of global geochemical cycles through killing microplankton, bacteria etc. Although viruses have no special "desire" to hurt the host cell, they very often are doing so just by selfishness and negligence.

It is well known that specific viruses can infect only specific organisms and can damage in these organisms only specific organs and tissues. An important factor contributing to this specific pathogenicity is the viral host range, i.e., the ability or inability to infect a given type of cells. Main factors controlling the host range are availability of appropriate receptors on the cell surface, appropriate intracellular milieu, and status of the innate immunity. Virus-induced damage to cells may be caused by many factors and among them by competition for resources and for cellular infrastructure. The outcome of infection also depends on the availability of viral anti-defensive mechanisms. In some instances, pathology may result from the hyperactive host defense.

Killing the host organism does not confer to the infecting virus any advantage. Rather, co-evolution of viruses and hosts should likely result in a kind of equilibrium. This principle may be illustrated by co-evolution of a rabbit-pathogenic virus and wild rabbits in Australia in the mid of the last century (Fenner and Ratcliffe, 1965). To control the enormous population of imported European rabbits, which became major pests for agricultural industry, it was decided to use the fibroma/ myxoma virus. This virus induces benign tumors in American rabbits but causes severe generalized lethal lesions in European ones. As expected, virus introduction resulted in mosquito-borne (summer) epizootics, with >99% of the infected rabbits dying in less than 2 weeks. Naturally occurring less virulent virus variants had more chances for overwintering, the circumstance resulting in selection of attenuated viruses. After a decade, the mortality of European rabbits in Australia caused by the evolved virus decreased roughly two-fold. Simultaneously, selection of resistant rabbits took place: mortality of such rabbits infected with the original virus decreased roughly fourfold.

On the other hand, newly emerging viral infections, caused often by viruses, which have been transferred to human populations from an animal reservoir, may exhibit a very high pathogenicity for humans. A remarkable example of this is the 1918 Spanish flu, which killed about 30–50 million people, or ~2% of those infected. In fact, influenza virus is in most cases a nonpathogenic or slightly pathogenic avian enteric virus. To infect a human, an avian flu virus should change its receptor specificity, which depends on the interaction of viral hemagglutinin (HA) with a cellular membrane glycoprotein receptor. Generally, it is sufficient to change only two amino acid residues in the avian HA to allow it to efficiently recognize the human receptor. Such a change in the host range may be achieved by either mutations in the avian HA or acquisition by an avian virus of the HA gene from human influenza virus as a result of genetic exchange (reassortment) between these viruses during mixed infections. Adaptation of flu viruses to humans may also require mutations in other viral genes. The severity of the infection depends not only on the efficiency of viral reproduction in the new host but also on the balance and interplay between host defense and viral counter-defense mechanisms.

Influenza virus populations are constantly changing. While undergoing immune pressure, they accumulate mutations, primarily in HA and neuraminidase (NA) ("antigenic drift"). Qualitatively new variants are emerging from time to time, primarily through reassortment ("antigenic shift"). The Spanish (1918) flu was just one of such "shifted" variants, as judged by the nucleotide sequence of its genomic RNA determined from archival formalin-treated paraffin-embedded as well as permafrost samples (Tumpey et al., 2005). It was established that the deadly virus was indeed of avian origin and that the "jump" into humans was associated with alterations in the HA and some other genes. The virus was reconstructed on the basis of known nucleotide sequence of its genome, and its pathogenic properties were confirmed and further investigated. The causative agents of the two next influenza pandemics, the "Asian" (1957) and "Hong-Kong" (1968) viruses, also resulted from reassortment between avian and human viruses. Currently newly emerging, highly pathogenic avian virus (H5N1) presents a potential threat of a new pandemic. Although cases of human-to-human transmission of this virus are so far extremely rare and only occur during close family contacts, the possibility of acquisition of ominous mutations cannot be ruled out.

The severe acute respiratory syndrome (SARS) is caused by another newly emerged virus. It belongs to coronaviruses. Representatives of this viral family are etiological agents of relatively mild respiratory (e.g., common cold), enteric, and some other diseases in a wide variety of animals and humans. Again, the very dangerous virus "came" to humans from an animal (in this case, bat) and the interspecies transmission involved a change in receptor recognition (Lau et al., 2005; Li et al., 2005).

The human immunodeficiency viruses (HIV) originated from simian immunodeficiency viruses (SIV). SIVs are viruses that infect some 40 different nonhuman primate species in sub-Saharan Africa. Primates naturally infected with SIV do not appear to develop immunodeficiency. The immediate precursor of HIV-1 was a virus, SIVcpz, which infects chimpanzees of the subspecies *Pan troglodytes troglodytes* in West Central Africa. Interestingly, SIVcpz strains have been transmitted from chimps to humans on at least three independent occasions, with the current HIV-1 group M pandemic strain (>60 million people infected, >20 million deaths) resulting from just one of these transmissions. The trans-species transmission appeared to require multiple adaptation events, such as changed transmission mode (from bite/wounds to sexual), adaptation to receptors and overcoming some post-entry barriers (Gao et al., 1999; Heeney et al., 2006). In human populations and individuals,

HIV is rapidly evolving by mutations and recombination, making antiviral therapy a very difficult job.

Major lessons derived from studies of emerging viral infections are as follows. New human pathogenic viruses are usually originating from relatively low pathogenic animal viruses by changing host range (crossing interspecies barrier). They may become highly pathogenic in humans, in particular, because they, as unknown invaders, are met with a hyper-reactive defense reaction leading to extensive host damage, whereas the viruses carelessly employ in full their anti-defensive tools. Further rapid evolutionary rates may contribute to difficulty in combating these infections. On the other hand, long-term evolution of such viruses may lead to a decrease in their pathogenicity and human evolution may probably result in an increased resistance to such viruses.

We may blame viruses for severe diseases and even should actively combat them but we should never forget that we owe to viruses very much for the existence of the live Nature and for our own existence.

References

Agol, V. I., 1974, Towards the system of viruses, *Biosystems*, **6**:113–132.

Baltimore, D., 1971, Expression of animal virus genomes, *Bacteriol. Rev.*, **35**:235–241.

Bamford, D. H., 2003, Do viruses form lineages across different domains of life? *Res. Microbiol.*, **154**:231–236.

Breitbart, M., and Rohwer, F., 2005, Here a virus, there a virus, everywhere the same virus? *Trends Microbiol.*, **13**:278–284.

Fenner, F., and Ratcliffe, F. N., 1965, Myxomatosis, Cambridge University Press, London and New York.

Forterre, P., 2006, Three RNA cells for ribosomal lineages and three DNA viruses to replicate their genomes: A hypothesis for the origin of cellular domain. *Proc. Natl. Acad. Sci. U. S. A.*, **103**:3669–3674.

Gao, F., Bailes, E., Robertson, D. L., Chen, Y., Rodenburg, C.M., Michael, S. F., Cummins, L. B., Arthur, L. O., Peeters, M., Shaw, G. M., Sharp, P. M., and Hahn, B. H., 1999, Origin of HIV-1 in the chimpanzee Pan troglodytes troglodytes. *Nature*, **397**:436–441.

Gmyl, A. P., Pilipenko, E. V., Maslova, S. V., Belov, G. A., and Agol, V. I., 1993, Functional and genetic plasticities of the poliovirus genome: Quasi-infectious RNAs modified in the 5′-untranslated region yield a variety of pseudorevertants. *J. Virol.*, **67**: 6309–6316.

Gmyl, A. P., Belousov, E. V., Maslova, S. V., Khitrina, E. V., and Agol V. I., 1999, Nonreplicative RNA recombination in poliovirus. *J. Virol.*, **73**:8958–8965.

Gmyl, A. P., Korshenko, S. A., Belousov, E. V., Khitrina, E. V., and Agol, V. I., 2003, Nonreplicative homologous RNA recombination: Promiscuous joining of RNA pieces? *RNA*, **9**:1221–1231.

Heeney, J. L., Dalgleish, A. G., and Weiss, R. A., 2006, Origins of HIV and the evolution of resistance to AIDS. *Science*, **313**:462–466.

Koonin, E. V., and Martin, W., 2005, On the origin of genomes and cells within inorganic compartments. *Trends Genet.*, **21**:647–654.

Koonin, E. V., Senkevich, T. G., and Dolja, V. V., 2006, The ancient virus world and evolution of cells, *Biology Direct.*, **1**:29.

Lau, S. K., Woo, P. C., Li, K. S., Huang, Y., Tsoi, H. W., Wong, B. H., Wong, S. S., Leung, S. Y., Chan, K. H., and Yuen, K. Y., 2005, Severe acute respiratory syndrome coronavirus-like virus in Chinese horseshoe bats. *Proc. Natl. Acad. Sci. U. S. A.*, **102**:14040–14045.

Li, W., Shi, Z., Yu, M., Ren, W., Smith, C., Epstein, J. H., Wang, H., Crameri, G., Hu, Z., Zhang, H., Zhang, J., McEachern, J., Field, H., Daszak, P., Eaton, B. T., Zhang, S., and Wang L.-F., 2005, Bats are natural reservoirs of SARS-like coronaviruses. *Science*, **310**:676–679.

Pfeiffer, J. K., and Kirkegaard, K., 2003, A single mutation in poliovirus RNA-dependent RNA polymerase confers resistance to mutagenic nucleotide analogs via increased fidelity, *Proc. Natl. Acad. Sci. U. S. A.*, **100**:7289–7294.

Pfeiffer, J. K., and Kirkegaard, K., 2005, Increased fidelity reduces poliovirus fitness and virulence under selective pressure in mice. *PLoS Pathog.*, **1**:e11.

Pilipenko, E. V., Gmyl, A. P., Maslova, S. V., Svitkin, Y. V., Sinyakov, A. N., and Agol, V. I., 1992, Prokaryotic-like cis element in the cap-independent internal initiation of translation on picornavirus RNA. *Cell*, **68**:119–131.

Pilipenko, E. V., Viktorova, E. G., Guest, S. T., Agol, V. I., and Roos, R. P., 2001, Cell-specific proteins regulate viral RNA translation and virus-induced disease. *EMBO J.*, **20**:6899–6908.

Suttle, C. A., 2005, Viruses in the sea. *Nature*, **437**:356–361.

Tumpey, T. M., Basler, C. F., Aguilar, P. V., Zeng, H., Solorzano, A., Swayne, D. E., Cox, N. J., Katz, J. M., Taubenberger, J. K., Palese, P., and Garcia-Sastre, A., 2005, Characterization of the reconstructed 1918 Spanish influenza pandemic virus. *Science*, **310**:77–80.

Vignuzzi, M., Stone, J. K., Arnold, J. J., Cameron, C. E., and Andino, R., 2006, Quasispecies diversity determines pathogenesis through cooperative interactions in a viral population. *Nature*, **439**:344–348.

Fokker-Planck and Chapman-Kolmogorov Equations for Ito Processes with Finite Memory

Joseph L. McCauley

Abstract The usual derivation of the Fokker-Planck partial differential eqn. (pde) assumes the Chapman-Kolmogorov equation for a Markov process [1, 2]. Starting instead with an Ito stochastic differential equation (sde), we argue that finitely many states of memory are allowed in Kolmogorov's two pdes, K1 (the backward time pde) and K2 (the Fokker-Planck pde), and show that a Chapman-Kolmogorov eqn. follows as well. We adapt Friedman's derivation [3] to emphasize that finite memory is not excluded. We then give an example of a Gaussian transition density with 1-state memory satisfying both K1, K2, and the Chapman-Kolmogorov eqns. We begin the paper by explaining the meaning of backward time diffusion, and end by using our interpretation to produce a very short proof that the Green function for the Black-Scholes pde describes a Martingale in the risk neutral discounted stock price.

Keywords Stochastic process, martingale, Ito process, stochastic differential eqn., memory, nonMarkov process, backward time diffusion, Fokker-Planck, Kolmogorov's partial differential eqns., Chapman-Kolmogorov eqn., Black-Scholes eqn.

Joseph L. McCauley
Physics Department
University of Houston
Houston, Tx. 77204-5005
jmccauley@uh.edu

and Senior Fellow
COBERA
Department of Economics
J.E. Cairnes Graduate School of Business and Public Policy
NUI Galway, Ireland

1 The Meaning of Kolmogorov's First PDE

Consider a diffusive process described by an Ito stochastic differential equation (sde) [3, 4] with or without finite memory in the drift and diffusion coefficients,

$$dx = R(x,t)dt + \sqrt{D(x,t)}dB(t) \tag{1}$$

where B(t) is the Wiener process [3, 4], so that $dB^2(t) = dt$ with probability one, $< dB(t)dB(s) >= 0$ if $s \neq t$. By finite memory, we mean explicitly a history of a finite nr. k of earlier states $(x_k, t_k; \ldots; x_1, t_1)$. This means that R and D may depend not only on the present state (x,t) but also on a specific history $(x_k, t_k; \ldots; x_1, t_1)$, so that the forward-time 2-point transition density $p_2(x, t+T : y, t)$ for the Ito process

$$x(t+T) = x(t) + \int_t^{t+T} R(x(s), s)ds + \int_t^{t+T} \sqrt{D(x(s), s)}dB(s) \tag{2}$$

also depends on history $(x_k, t_k; \ldots; x_1, t_1)$, where $t \geq t - T \geq t_k \geq \ldots \geq t_1$. First, however, we derive the pde for the backward time 2-point transition density.

Consider a measurable, twice differentiable dynamical variable A(x,t). The sde for A is (by Ito's lemma [3–5])

$$dA = \left(\frac{\partial A}{\partial t} + R\frac{\partial A}{\partial x} + \frac{D}{2}\frac{\partial^2 A}{\partial x^2}\right)dt + \sqrt{D}\frac{\partial A}{\partial x}dB \tag{3}$$

so that

$$\begin{aligned} A(x(t+T), t+T) = A(x(t), t) &+ \int_t^{t+T} \left(\frac{\partial A(x(s), s)}{\partial t} + R\frac{\partial A}{\partial x} + \frac{D}{2}\frac{\partial^2 A}{\partial x^2}\right) ds \\ &+ \int_t^{t+T} \sqrt{D(x(s), s)}\frac{\partial A(x(s), s)}{\partial x}dB(s) \end{aligned} \tag{4}$$

A martingale is defined by the conditional average $< A(x, t+T) >_c = A(x,t)$ [4, 5] where a backward in time average is indicated[1] and where $< x(t+T) >_c = x(t) + \int < R >_c ds$. We want to obtain the generator for the backward time transition probability density $p^+(x, t : y, t+T)$. This can be done in either of two ways. First, with $T > 0$ we have by definition that

$$A(x,t) = \int p^+(x, t : y, t+T) A(y, t+T) dy \tag{5}$$

[1] A martingale forecast represents a 'fair game' condition.

backward in time and $A(y,t+T)$ is to be specified as a forward in time initial condition. For analytic functions A we have

$$A(x,t) \approx A(x,t+T) + \int p^{+}(x,t:y,t+T)\left((y-x)\frac{\partial A(x,t+T)}{\partial x}\right.$$
$$\left. + (y-x)^2\frac{\partial^2 A}{\partial x^2}\right) dy + \ldots \quad (6)$$

so that with T vanishing, and using the usual definition of drift and diffusion coefficients [1], we obtain the backward time diffusion pde [3, 4], the generator of the martingale,

$$0 = \frac{\partial A(x,t)}{\partial t} + R(x,t)\frac{\partial A(x,t)}{\partial x} + \frac{D(x,t)}{2}\frac{\partial^2 A(x,t)}{\partial x^2} \quad (7)$$

if the moments $\int (y-x)^n p^{+} dy$ vanish fast enough with T for $n \geq 3$. If the transition density depends on a *finite* history of exactly k earlier states, $p^{+}(x,t:y,s) = p^{+}{}_{n}(x,t:y,s;x_k,t_k;\ldots;x_1,t_1)$ with $k = n-2$, then that history appears in the drift and diffusion coefficients as well, e.g.,

$$D(x,t;x_k,t_k,\ldots,x_1,t_1) \approx \frac{1}{T}\int (y-x)^2 p_{k+2}{}^{+}(y,t:x,t-T;x_k,t_k;\ldots;x_1,t_1)dy \quad (8)$$

as T vanishes.

Second, note that the backward time pde (7) follows directly from (4) simply by setting the drift term equal to zero, yielding a martingale

$$A(x(t+T),t+T) = A(x(t),t) + \int_{t}^{t+T} \sqrt{D(x(s),s)}\frac{\partial A(x(s),s)}{\partial x} dB(s) \quad (9)$$

In this second and more general derivation no assumption is made or needed either of the moments $\int (y-x)^n p^{+} dy$ vanishing fast enough with T for $n \geq 3$, or of analyticity of A(x,t).

We've made no assumption that A is positive. I.e., A is generally not a 1-point probability density, A(x,t) is simply any martingale, and an infinity of martingales can be so constructed depending on the choice of forward time initial conditions specified on A (an initial value or boundary value problem backward in time is solved [3, 5]). By (5) the required transition density is the Green function of (7),

$$0 = \frac{\partial g^{+}(x,t:y,s)}{\partial t} + R(x,t)\frac{\partial g^{+}(x,t:y,s)}{\partial x} + \frac{D(x,t)}{2}\frac{\partial^2 g^{+}(x,t:y,s)}{\partial x^2} \quad (10)$$

where $g^{+}(x,t:y,t) = \delta(x-y)$. I.e., $p^{+}(x,t:y,s) = g^{+}(x,t:y,s)$ with $t < s$. The conditions under which g^{+} exists, is unique and nonnegative definite are stated in Friedman [3]. Equation (10) is called Kolmogorov's first pde (K1) [1].

What does K1 mean? Simply that martingales can be constructed via Ito's lemma.

2 The Fokker-Planck PDE with Finite Memory

Consider next *any* measurable twice-differentiable dynamical variable A(x(t)). A(x) is *not* assumed to be a martingale. The time evolution of A is given by Ito's lemma [6, 7]

$$dA = \left(R\frac{\partial A}{\partial x} + \frac{D}{2}\frac{\partial^2 A}{\partial x^2}\right)dt + \sqrt{D}\frac{\partial A}{\partial x}dB \tag{11}$$

We can calculate the conditional average of A, conditioned on $x(t_o) = x_o$ at time t_o in $x(t) = x_o + \int R(x,s)ds + \int \sqrt{D}(x,s)dB(s)$, forward in time if we know the transition density $p_2(x,t : x_o, t_o))$ forward in time,

$$\langle A(x(t))\rangle = \int p_2(x,t : x_o, t_o)A(x)dx \tag{12}$$

Note that this is not the rule for the time evolution of a 1-point probability density. From

$$\frac{d\langle A(x(t))\rangle}{dt} = \int \frac{\partial p_2(x,t : x_o, t_o)}{\partial t}A(x)dx \tag{13}$$

and using

$$\langle dA\rangle = \left(\left\langle R\frac{\partial A}{\partial x}\right\rangle + \left\langle\frac{D}{2}\frac{\partial^2 A}{\partial x^2}\right\rangle\right)dt \tag{14}$$

with $< dA > /dt$ defined by (13), we obtain from (14), after integrating twice by parts and assuming that the boundary terms vanish,

$$\int dxA(x)\left[\frac{\partial p_2}{\partial t} + \frac{\partial(Rp_2)}{\partial x} - \frac{1}{2}\frac{\partial^2(Dp_2)}{\partial x^2}\right] = 0 \tag{15}$$

so that the transition density is the Green function of the Fokker-Planck pde [1–4], or Kolmogorov's second pde (K2)

$$\frac{\partial p_2}{\partial t} = -\frac{\partial(Rp_2)}{\partial x} + \frac{1}{2}\frac{\partial^2(Dp_2)}{\partial x^2} \tag{16}$$

Since p_2 is a transition density we also have the 2-point density $f_2(x,t;y,s) = p_2(x,t : y,s)p_1(y,s)$ where the 1-point density $f_1 = p_1$ satisfies

$$p_1(x,t) = \int f_2(x,t;y,s)dy = \int p_2(x,t : y,s)p_1(y,s)dy \tag{17}$$

and so satisfies the same pde (16) as does p_2 but with an arbitrary initial condition $p_1(x,t_1) = f(x)$. Note the difference with (12). So far, no Markovian assumption was made.

In particular, no assumption was made that R, D, and hence p_2, are independent of memory of an initial state, or of finitely many earlier states. If there is memory, e.g., if $p_1(x,t_o) = u(x)$ and if $D = D(x,t;x_o,t_o)$ depends on one initial state

$x_o = \int xu(x)dx$, then due to memory in $p_1(x,t)$ [8],

$$p_2(x_3,t_3 : x_2,t_2) = \frac{\int p_3(x_3,t_3 : x_2,t_2;x_1,t_1)p_2(x_2,t_2 : x_1,t_1)p_1(x_1,t_1)dx_1}{\int p_2(x_2,t_2 : x_1,t_1)p_1(x_1,t_1)dx_1} \tag{18}$$

then by the 2-point transition density we must understand that $p_2(x,t : y,s) = p_3(x,t : y,s;x_o,t_1)$. That is, in the simplest case p_3 is required to describe the stochastic process. Memory appears in (18) if, e.g., at time t_o $f(x) = \delta(x - x_o)$ with $x_o \neq 0$ [8]. The main idea is that we are dealing quite generally with Ito sdes and corresponding pdes for transition densities with memory of a finite nr. n-2 of states, so that the 2-point transition density is $p(x,t : y,s) = p_n(x,t : y,s;x_{n-2},t_{n-2};\ldots;x_1,t_1)$ depending on $n - 2$ earlier states.

Now, for the case where A(x(t)) *is* a martingale (requiring that the drift term in (11) vanishes), then (12) must yield

$$\langle A\rangle_t = \int p(x,t : x_o,t_o)A(x)dx = A(x_o) \tag{19}$$

and since (19) cannot differ from (5) if the theory is to make any sense, then there must be a connection between the backward and forward time transition densities p^+ and p_2. Comparing (19) with (5) we see that p^+ and p_2 must be adjoints. For a Markov process it's very easy to use the Chapman-Kolmogorov eqn. to derive both [1] the Fokker-Planck and Kolmogorov's backward time pde, and then prove that the Green function of K1 is the adjoint of the Green function of K2, but we will avoid making any Markovian assumption in order to permit finite memory in the formalism. In particular, we have not and will not assume in advance that a Chapman-Kolmogorov eqn. holds, but will next explain why that eqn. follows, even with finite memory. Then, in part 4, we'll derive the Chapman-Kolmogorov (21) below from memory dependent pdes K1 and K2.

3 The Chapman-Kolmogorov Equation for Finite Memory

That a Chapman-Kolmogorov eqn. should hold for finitely many states of memory follows from standard definitions of conditional probability densities. With an unstated, even infinite, number of states in memory the history-dependent 2-point transition densities obey the hierarchy

$$\begin{aligned} p_{k-1}(x_k,t_k|x_{k-2},t_{k-2};\ldots;x_1,t_1) = \int dx_{k-1}p_k(x_k,t_k|x_{k-1},t_{k-1};\ldots;x_1,t_1) \\ p_{k-1}(x_{k-1},t_{k-1}|x_{k-2},t_{k-2};\ldots;x_1,t_1) \end{aligned} \tag{20}$$

For fractional Brownian motion (fBm), e.g., there is no reason to expect this hierarchy to truncate. But consider processes where the memory is finite and of number $n - 2$, so that $p_k = p_n$ for all $k \geq n$. Then from (20) we obtain the

Chapman-Kolmogorov eqn. in the form [9]

$$p_n(x_n,t_n|x_{n-1},t_{n-1};\ldots;x_1,t_1) = \int dy p_n(x_n,t_n|y,s;x_{n-2},t_{n-2};\ldots;x_1,t_1)$$
$$p_n(y,s|x_{n-1},t_{n-1};\ldots;x_1,t_1) \tag{21}$$

for a process with finite memory. Next, for completeness, we will take a step backward and show that the pde K1 for an Ito process (1) with finite memory in R and/or D implies both the Fokker-Planck pde and the Chapman-Kolmogorov eqn.

4 Ito Implies K1 and K2 Implies Chapman-Kolmogorov, Even with Finite Memory

Consider the linear operators

$$L^+ = \partial/\partial t + R(x,t)\partial/\partial x + (D(x,t)/2)\partial^2/\partial x^2 \tag{22}$$

and

$$Lu = -\partial u/\partial t + \partial(R(x,t)u)/\partial x - \partial^2(D(x,t)u/2)/\partial x^2 \tag{23}$$

acting on a function space of measurable, twice (not necessarily continuously) differentiable functions satisfying boundary conditions at $t=\infty$, and at $x=-\infty$ and $x=\infty$ to be indicated below. Both operators followed superficially independently above from the Ito process (1), but we can start with (22) and then obtain (23) via

$$uL^+v - vLu = \frac{\partial}{\partial t}(uv) + \frac{\partial}{\partial x}\left(vRu + \frac{1}{2}uD\frac{\partial v}{\partial x} - v\frac{1}{2}\frac{\partial uD}{\partial x}\right) \tag{24}$$

which is a form of Green's identity (see also [3], but where the operator L is studied in standard elliptic rather than in Fokker-Planck form). With suitable boundary conditions on u,v [4] then L and L^+ are adjoints of each other:

$$\int_0^\infty dt \int_{-\infty}^\infty (vLu - uL^+v)dx = 0 \tag{25}$$

Starting with an Ito process (1) and K1, we have deduced K2. No Markovian assumption has been made. Again, the formal conditions under which (25) holds are stated in Friedman [3].

Next, let $g^+(x,t:x,t)$ denote the Green function of K1, $L^+g^+=0$, and let $g(x,t:x,t)$ denote the Green function of K2, $Lg=0$. Let $t<s<t$ and assume also that $t+e<s<t-e$, which avoids sitting on top of a delta function. Integrating (24) over y from $-\infty$ to ∞ and over s from $t+e$ to $t-e$ with the choices $v(y,s)=g^+(y,s:x,t)$ and $u(y,s)=g(y,s:x,t)$, we obtain [3]

$$\int g(y,t-\varepsilon:\xi,\tau)g^+(y,t-\varepsilon:x,t)dy = \int g(y,\tau+\varepsilon:\xi,\tau)g^+(\tau+\varepsilon:x,t)dy \tag{26}$$

With e vanishing and using $g(y,t:x,t) = d(y-x)$, $g^+(y,t:x,t) = \delta(y-x)$, we obtain the adjoint condition for the Green functions

$$g(x,t:\xi,\tau) = g^+(\xi,\tau:x,t) \tag{27}$$

Next, apply the same argument but with times $t \leq t'' \leq t' \leq t$ to obtain (instead of (26))

$$\int g(y,t':\xi,\tau)g(x,t:y,t')dy = \int g(y,t'':\xi,\tau)g(x,t:y,t'')dy \tag{28}$$

If we let t'' approach ṭ then we obtain the Chapman-Kolmogorov eqn.

$$g(x,t:\xi,\tau) = \int g(x,t:y,t')g(y,t':\xi,\tau)dy \tag{29}$$

again, *without having made any Markovian assumption.* The considerations of parts 2 and 3 tell us that we must restrict to transition densities depending at most on only finitely many states in memory.

Summarizing, beginning with the Ito sde (1) and obtaining K1 (10) we've deduced K2 and finally the Chapman-Kolmogorov eqn. The derivation follows Friedman's [3] where a Markov process was claimed, but we see that nowhere was the assumption of a Markov process either used or needed. The implication is that, with suitable boundary conditions on Green functions, an Ito sde implies both K1 and K2 and the Chapman-Kolmogorov eqn., even with finite memory. (Equation (21) may make no sense even for countably infinitely many states in memory, and demonstrably does not hold for non-Ito processes like fBm [7, 10]).

To show that this new formalism is not vacuous, we now provide a simple example. We provide no example for variable diffusion D(x,t) where the (x,t) dependence is not separable, because even for the scaling class of models [6] we do not yet know how to calculate a model green function analytically.

5 A Gaussian Process with 1-State Memory

Consider first the 2-point transition density for an arbitrary Gaussian process in the form [8]

$$p(x,t:y,s) = \frac{1}{\sqrt{2\pi K(t,s)}} e^{-(x-m(t,s)y-g(t,s))^2/2K(t,s)} \tag{30}$$

Until the pair correlation function $\langle x(t)x(s)\rangle$ a m(t,s) is specified, no particular process is indicated by (30). Processes as wildly different and unrelated as fBm [10], scaling Markov processes [10], and Ornstein-Uhlenbeck processes [11] are allowed. Depending on the pair correlation function $\langle x(t)x(s)\rangle$, memory, including long time memory, may or may not appear. To obtain fBm, e.g., $g = 0$ and $\langle x(s)x(t)\rangle$ must reflect the condition for stationary increments [10], which differs strongly from a condition of time translational invariance whereby m, g, and K may depend on (s,t) only in the form s-t. Fortunately, Hänggi and Thomas [8] have stated the conditions for a Gaussian process (30) to satisfy a Chapman-Kolmogorov eqn., namely,

$$\begin{aligned} m(t,t_1) &= m(t,s)m(s,t_1) \\ g(t,t_1) &= g(t,s)+m(t,s)g(s,t_1) \\ K(t,t_1) &= K(t,s)+m^2(t,s)K(s,t_1) \end{aligned} \tag{31}$$

Actually, Hänggi and Thomas stated in [8] that (31) is the condition for a *Markov* process, but we will show that the Chapman-Kolmogorov condition (31) is satisfied by at least one Gaussian process with memory.

Consider next the 1-point density $p_1(x,t)$ for a specific Ito process with simple memory in the drift coefficient, the Shimizu-Yamato model [9, 12]

$$\frac{\partial p_1}{\partial t} = \frac{\partial}{\partial x}((\gamma+\kappa)x - \kappa\langle x(t)\rangle + \frac{Q}{2}\frac{\partial}{\partial x})p_1 \tag{32}$$

with initial data $p(x,t_o) = f(x)$ and with $<x(t)>= \int xp_1(x,t)dx$. The parameter Q is the diffusion constant. Since the drift coefficient in (1) is $R = -(g+k)x + k < x(t) >$, and since [13]

$$\frac{d\langle x\rangle}{dt} = \langle R\rangle = -\gamma\langle x\rangle \tag{33}$$

we obtain

$$\langle x(t)\rangle = x_o e^{-\gamma(t-t_o)} \tag{34}$$

where

$$x_o = \int xf(x)dx \tag{35}$$

This provides us with a drift coefficient with initial state memory,

$$R(x,t;x_o,t_o) = -(\gamma+\kappa)x + \kappa x_o e^{-\gamma(t-t_o)} \tag{36}$$

Because $g \neq 0$ the memory cannot be eliminated via a simple coordinate transformation $z = x - <x>$.

The Fokker-Planck pde for the transition density $p_2(x,t:y,s;x_o,t_o)$ is

$$\frac{\partial p_2}{\partial t} = \frac{\partial}{\partial x}\left((\gamma+\kappa)x - \kappa x_o e^{-\gamma(t-t_o)} + \frac{Q}{2}\frac{\partial}{\partial x}\right)p_2 \tag{37}$$

with $p_2(x,t:y,t;x_o,t_o) = d(x-y)$. The solution is a Gaussian (30) with 1-state memory where

$$\begin{aligned} m(t,s) &= e^{-(\gamma+\kappa)(t-s)} \\ K(t,s) &= \frac{Q}{\gamma+\kappa}(1-e^{-2(\gamma+\kappa)(t-s)}) \\ g(t,s) &= x_o(e^{-\gamma(t-t_o)} - e^{-(\gamma+\kappa)t+\gamma t_o+\kappa s}) \end{aligned} \tag{38}$$

An easy calculation shows that the Chapman-Kolmogorov conditions (31) are satisfied with finite memory (x_o,t_o). Furthermore, $p^+(y,s:x,t;x_o,t_o) =$

$p_2(x,t:y,s;x_o,t_o)$ satisfies the backward time diffusion pde K1 in the variables (y,s),

$$0=\frac{\partial p^+}{\partial s}+R(y,s;x_o,t_o)\frac{\partial p^+}{\partial y}+\frac{Q}{2}\frac{\partial^2 p^+}{\partial y^2} \tag{39}$$

with drift coefficient

$$R(y,s;x_o,t_o)=-(\gamma+\kappa)x+\kappa x_o e^{-\gamma(s-t_o)} \tag{40}$$

This shows that backward time diffusion makes sense in the face of memory. The memory simply yields $p^+(y,t_o:x_o,t_o;x_o,t_o)=\delta(y-x_o)$.

6 Black-Scholes from a Different Standpoint

Recapitulating, we understand the meaning of backward time diffusion qualitatively: we can construct martingales from an Ito process via Ito's lemma by setting the drift coefficient equal to zero, yielding K1 (see Steele [5] for simple but instructive martingales that can be constructed by solving (7) for various different initial and boundary conditions). This insight allows us to prove directly from the risk neutral hedge, the so-called delta hedge [13], that the Black-Scholes pde describes a martingale in the risk neutral discounted 'stock' price. We begin with the sde for the stock price p(t),

$$dp=\mu p dt+p\sqrt{d(p,t)}dB \tag{41}$$

where m is the unreliably known or estimated 'interest rate' on the stock. In the delta hedge strategy, w(p,t) is the option price and satisfies the Black-Scholes pde [13]

$$rpw(p,t)=\frac{\partial w(p,t)}{\partial t}+rp\frac{\partial w(p,t)}{\partial p}+\frac{p^2 d(p,t)}{2}\frac{\partial^2 w(p,t)}{\partial p^2} \tag{42}$$

where r is the risk free interest rate (the interest rate on a bank deposit, money market fund, or CD). With $v=we^{r(t-T)}$, where T is the expiration time of a 'European' option,

$$0=\frac{\partial v(p,t)}{\partial t}+rp\frac{\partial v(p,t)}{\partial p}+\frac{p^2 d(p,t)}{2}\frac{\partial^2 v(p,t)}{\partial p^2} \tag{43}$$

By (6), v defined by (43) is a martingale so that w(p,t) describes a martingale in the risk neutral discounted option price. That is, this model predicts a theoretically 'fair' option price, and corresponds to a stock price S(t) where the interest rate is r,

$$dS=rSdt+S\sqrt{d(S,t)}dB \tag{44}$$

See [14] for a longer proof that the Green function for the Black-Scholes pde (42) describes a martingale in the risk neutral discounted stock price. From our standpoint, the Black-Scholes pde is simply a standard equation of martingale construction for Ito processes, and we see that finite memory may indeed appear in a 'fair' option price.

For purists who are allergic to econophysics, one can derive the Black-Scholes pde (42) without mentioning the financial motivation. Starting with the process (41), which reduces to the lognormal process if $d(p,t) = d(t)$, define a new stochastic process $P = -w(p,t) + Dp$ and impose the condition that P evolves locally deterministically (i.e., require that dP/dt has no noise term). This requires the choice $D = w'$. One can generalize this to sdes other than (41). I.e., the problem solved by Black-Scholes is: given a stochastic process, how can we construct a new stochastic process such that the sum of the two processes is locally deterministic. In finance, one wants the deterministic process to evolve in time with the bank interest rate r, $dP/dt = rP$, in order to construct a hypothetically arbitrage free condition.

We end with two remarks. First, Friedman [3] shows that the Chapman-Kolmogorov eqn. is not restricted to K1 and its adjoint the Fokker-Planck pde, but holds more generally for Green functions of pdes of the type

$$L^{+}v = \partial v/\partial t + c(x,t)v + R(x,t)\partial v/\partial x + (D(x,t)/2)\partial^{2}v/\partial x^{2} = 0 \qquad (45)$$

and its adjoint. With $c = -R = rx$ and $D = x^{2}d(x,t)$, where $x = p$ is in this case the stock price, we obtain the Black-Scholes pde (42). That the Green function for the Black-Scholes pde obeys the Chapman-Kolmogorov eqn. is surprising. More generally, $c(x,t) = R(x,t)$ defines the Black-Scholes class of equations for variable diffusion coefficients and variable interest rates.

Initial value problems of (45), where u(x,T) is specified at a forward time $T > t$, are solved by a Martingale construction that results in the Feynman-Kac formula [3]. Defining $M(s) = v(x,s)I(s)$, with dv(x,s) given by Ito's lemma and using (45) we obtain

$$\begin{aligned} dM = dvI + vdI &= -c(x,s)v(x,s)ds + v(x,s)dI(s) \\ &\quad + \sqrt{D(x(s),s)}\frac{\partial v}{\partial x}I(s)dB(s) \end{aligned} \qquad (46)$$

We obtain a martingale $M(s) = v(x,s)$ with the choice

$$I(s) = e^{-\int_{s}^{t} c(x(q),q)dq} \qquad (47)$$

so that the solution of (45) is given by the martingale condition $M(t) = \langle M(T) \rangle$,

$$v(x,t) = \left\langle v(x(T),T)e^{\int_{t}^{T} c(x(s),s)ds} \right\rangle \qquad (48)$$

where the Feynman-Katz average (48) at time T is calculated using the Green function $g^{+}(x,t : x(s),s)$ of (45) with $c = 0$, i.e., the Green function of K1. This martingale construction for solutions of Black-Scholes type pdes (45) is given in [5][2] using

[2] In [5], eqns. (15.25) and (15.27) are inconsistent with each other, (15.25) cannot be obtained from (15.27) by a shift of coordinate origin because the x-dependent drift and diffusion coefficients break translation invariance. A careful treatment of solving elliptic and parabolic pdes by running an Ito process is provided by Friedman [3].

unnecessarily complicated notation, and without the explanation of the connection of the Black-Scholes pde with K1, K2, and the Chapman-Kolmogorov eqn. It was derived by Friedman [3] over twenty years before it was advertised in the financial economics literature by Duffie [16].

Second, although 'memory' (or 'after effect') is never mentioned in the text [3], according to Friedman's definition processes with memory should labeled as 'Markov Processes' so long as the Chapman-Kolmogorov eqn. is satisfied. His definition of a Markov process (p. 18, ref. [3]), stated here in terms of transition densities, is that (1) there exists a (Borel) measurable transition density $p(x,t:y,s) \geq 0$, (2) that p(x,t:y,s) is a (probability) measure, and (3) that p(x,t:y,s) satisfies the Chapman-Kolmogorov eqn. By this definition the Shimizu-Yamada model is Markovian. However, this classification contradicts the standard definition of 'Markov' as a process 'without aftereffect' [1, 2], without history dependence [4, 15]. A Markov process is typically defined as a process whereby the time evolution of the transition density p(x,t:y,s) is fixed by specifying exactly one earlier state (y,s), $s < t$. For experts like Feller [17] and Doob [18] as well, the Chapman-Kolmogorov eqn. is a necessary but insufficient condition for a Markov process. The strength of Friedman's text is that it teaches us classes of diffusive nonMarkovian systems that satisfy that condition. Feller's example of a nonMarkov process satisfying the Chapman-Kolmogorov eqn. is discrete [17].

Acknowledgements JMC thanks Enrico Scalas, Harry Thomas, Gemunu Gunaratne, and Kevin Bassler for many pleasant and very stimulating discussions via email about various points. Without HT having sent us references [8, 11], and without ES having pointed us to Doob's assertion that the Chapman-Kolmogorov eqn. is a necessary but insufficient condition for a Markov process [17], which led to Feller's paper [16], this paper would not have been possible.

References

1. B.V. Gnedenko, *The Theory of Probability*, tr. by B.D. Seckler (Chelsea, NY, 1967).
2. R.L. Stratonovich. *Topics in the Theory of Random Noise*, tr. by R.A. Silverman (Gordon & Breach, NY, 1963).
3. A. Friedman, *Stochastic Differential Equations and Applications* (Academic, NY, 1975).
4. L. Arnold, *Stochastic Differential Equations* (Krieger, Malabar, 1992).
5. J.M. Steele, *Stochastic Calculus and Financial Applications* (Springer-Verlag, NY, 2000).
6. K.E. Bassler, G.H. Gunaratne, & J.L. McCauley, *Hurst Exponents, Markov Processes, and Nonlinear Diffusion Equations, Physica A* **369**: 343 (2006).
7. J.L. McCauley, G.H. Gunaratne, & K.E. Bassler, *Martingales, Detrending Data, and the Efficient Market Hypothesis, Physica A* **387**: 3916–3920 (2008).
8. P. Hänggi & H. Thomas, *Time Evolution, Correlations, and Linear Response of Non-Markov Processes, Zeitschr. Für Physik* **B26**: 85 (1977).
9. J.L. McCauley, *Markov vs. nonMarkovian Processes: A Comment on the Paper 'Stochastic Feedback, Nonlinear Families of Markov Processes, and Nonlinear Fokker-Planck Equations*, by T.D. Frank, *Physica A* **382**: 445–452 (2007).
10. J.L. McCauley, G.H. Gunaratne, & K.E. Bassler, *Hurst Exponents, Markov Processes, and Fractional Brownian Motion, Physica A* **379**: 1–9 (2007).

11. P. Hänggi, H. Thomas, H. Grabert, & P. Talkner, *Note on time Evolution of Non-Markov Processes, J. Stat. Phys.* **18**: 155 (1978).
12. T.D. Frank, *Stochastic Feedback, Nonlinear Families of Markov Processes, and Nonlinear Fokker-Planck Equations, Physica A* **331**: 391 (2004).
13. J.L. McCauley, *Dynamics of Markets: Econophysics and Finance* (Cambridge, Cambridge,2004).
14. J.L. McCauley, G.H. Gunaratne, & K.E. Bassler, *Martingale Option Pricing, Physica A* **380**: 351–356 (2007).
15. M.C. Wang & G.E. Uhlenbeck in *Selected Papers on Noise and Stochastic Processes*, ed. by N. Wax (Dover, NY, 1954).
16. D. Duffie, *An Extension of the Black-Scholes Model of Security Valuation, J. Econ. Theory* **46**,194, 1988.
17. W. Feller, *The Annals of Math. Statistics* **30**, No. 4, 1252, 1959.
18. J.L. Snell, *A Conversation with Joe Doob*, http://www.dartmouth.edu/~chance/Doob/conversation.html; *Statistical Science* **12**, No. 4, 301, 1997.

Evolution of FX Markets via Globalization of Capital

Joseph L. McCauley

Abstract This paper is about money, and why today's foreign exchange (FX) markets are unstable. According to the literature [1], FX markets were fundamentally different before and after WW I. Any attempt to discuss this topic within standard economic theory necessarily fails because money/liquidity/uncertainty is completely excluded from that theory [2]. Fortunately, our market dynamics models adequately serve our purpose. Eichengreen [1] has presented a stimulating history of the evolution of FX markets from the gold standard of the late nineteenth century through the Bretton Woods Agreement (post WWII–1971) and later the floating currencies of our present market deregulation era (1971–present). He asserts a change from stability to instability over the time interval of WWI. Making his argument precise, we describe how speculators could have made money systematically from a market in statistical equilibrium. The present era normal liquid FX markets are in contrast very hard, to a first approximation impossible, to beat, and consequently are described as 'martingales'. The ideas of martingales and options/hedging were irrelevant in the pre-WWI era. I end my historical discussion with the empirical evidence for the stochastic model that describes FX market dynamics quantitatively accurately during the last 7–17 years [3].

Keywords FX market instability, martingales, options and hedging

Joseph L. McCauley
Physics Department
University of Houston
Houston, TX 77204-5005
jmccauley@uh.edu

1 Pre-WWI Markets

1.1 The Gold Standard

The gold standard became widely accepted around 1890. But even then the 'quantity theory of money' (and especially conservation of money) did not apply. With a strict gold standard, and no new mining production or coinage of existing gold, money would be conserved in the absence of credit. But credit has played a strong role in finance and economics since at least the renaissance in Europe, and dominates all markets today: via any credit card purchase money is created with the tap of a computer key. Today, the 'amount of money' in the world is an ambiguous notion.

The main theme of the next section is that before WWI stable currency values were maintained by the threat of central bank intervention. This led speculators to bid up a weak currency with the expectation of a profit, and thereby strengthened the currency via a self-fulfilling process: speculation in that era tended to stabilize FX markets. After WWI the central bank threat no longer carried sufficient weight, and so FX markets became unstable: weak currencies were bid lower by speculators. The historic reasons for the change are discussed. This paper can be understood as adding equations and some interesting anecdotes to Eichengreen's history [1] of the globalization of capital.

1.2 How FX Stability Worked on the Gold Standard

Adhering to a gold standard means tight money. Credit cards don't exist, credit is hard to get. Banks in the gold standard era did not easily make loans for consumption. The level of economic activity was correspondingly much less than in our present era, and the price level was correspondingly lower. Markets were relatively illiquid, meaning that items were not frequently traded. Economic expansion/consumption requires easy money via credit. We begin with the pre-WWI era in which gold, not credit, was not the overwhelmingly dominant factor in finance.

In the nineteenth century there had been competition between silver and gold to set the monetary standard. For a while both metals were used, but the bimetallic monetary standard was largely eliminated in the late nineteenth century when silver became cheap in terms of gold. Great Britain's reliance on silver was an exception. According to Eichengreen's history of the gold standard and its eventual replacement as international standard by the post WWII Dollar under the Bretton Woods Agreement [1], *we can infer that there was a fundamental shift in the foreign exchange (FX) noise distribution after WWI.* Until the Great Depression, the Western countries followed the gold standard within bounds set by central banks.

The main job of western central banks and parliaments was apparently seen as keeping the national currency from falling outside chosen gold standard bands.

Eichengreen claims that before WWI the currency speculators, confident that governments could be relied on eventually to react to maintain the gold-value of the currency, bid a weak currency up, expecting a profit in the long run. Unfortunately, the older FX data are too poor to test this idea but, if true, that would mean that financial markets were asymptotically stationary in the pre-WWI era. This is interesting, because we know empirically that FX markets at least since 1958 are dynamically unstable (are nonstationary). Eichengreen's claim is that the onset of the instability coincided with social changes due to WWI.

Here's how the pre-WWI FX market presumably worked. Imagine a Dollar equivalent to 25.9 grains gold. Take the Reichsmark (RM) (the forerunner of the Deutschmark and then the Euro) as the foreign currency of interest, and focus on trade with Germany. Assume that credit (money creation without gold backing) doesn't change the money supply significantly inside the U.S. A trade deficit means too many Dollars outside the country, therefore too few inside the country so that economic activity within the U.S. falls. I.e., the trade deficit reduces liquidity inside a country on the gold standard (meaning deflation, unless more money is printed). Less money means lower prices (deflation), so the trade deficit is eventually reversed via cheaper exports. The latter brings money back into the country. This increases the Dollar against the RM without the need for a devaluation of the weak Dollar by the central bank. Incidentally, the central bank in the U.S. was created late, in 1913. By reducing the money supply (weakening demand further), a central bank could speed up this process. We can describe this stabilization process mathematically.

Consider the logarithmic return [2] $x(t) = \ln p(t)/p_c$ where p is the price of one currency in units of another (e.g., the Reichsmark in 1913, or the Euro today, in Dollars), and p_c is a reference price to be explained below.

In a *stationary* process, the 1-point returns density f(x,t) must be time-independent: the average return, the variance, and all other moments of the 1-point distribution are therefore constants. A market in statistical equilibrium reflects a stationary process x(t). We can easily model an asymptotically stationary market. With the usual stochastic supply–demand equation

$$dp = rpdt + \sigma_1\, pdB(t) \tag{1}$$

where B(t) is the Wiener process (dB/dt is white noise), we obtain (via Ito calculus)

$$dx = (r - \sigma_1{}^2/2)dt + \sigma_1 dB(t) \tag{2}$$

Let $R = r - s_1{}^2/2$ denote the expected return, $<x>= R$. For FX markets we know empirically [3] that $R \approx 0$. If $-\infty < x < \infty$, then (2) is the lognormal model introduced by Osborne in 1958 and used by Black and Scholes in 1973. Here, and in our more general models, the reference price p_c is the price at which the returns distribution peaks and defines what we call "value" [4]. "Value", or most probable price, is the price agreed on by the largest fraction of speculators. The speculators' behavior generates the noise distribution, which in the case of (2) is white. But central bank intervention means that $-\infty < x < \infty$ is the wrong assumption.

With the pre-WWI Dollar supported within a gold band $b_1 < x < b_2$, we can set the equilibrium probability density $f(x) = $ constant except at the boundaries and then we obtain an approach to statistical equilibrium: f(x,t) approaches f(x) as t increases (see Stratonovich [5] for the mathematical details). I.e., the market is asymptotically stationary. Here's how speculators could systematically suck money out of a stationary market [4]. Consider the price distribution $g(p,t) = f(x,t)dx/dp$ with price variance $\sigma_p{}^2$. One buys, e.g., if $p < p_G + \sigma_p$, and one sells if $p > p_G - \sigma_p$. *Such fluctuations are guaranteed because a stationary process is ergodic* [6], and the first passage time for a specific fluctuation can easily be calculated [5, 7]. So we understand how speculators could systematically have made money, as Eichengreen claims, *with little risk* in the pre-WWI era.

In a stationary market, returns merely fluctuate about the statistical equilibrium return $x = 0$, with constant variance. The 2-point and all other higher order densities are time translationally invariant. E.g., the conditional 2-point density obeys $p_2(y,s:x,t) = p_2(y,s\text{-}t:x,0)$. Time translational invariance alone is a necessary but not sufficient condition for statistical equilibrium [8], because time translational invariance does not guarantee the existence of a normalizable time-independent 1-point density f(x) [9]. This point is not appreciated in the economics and finance literature, where stationarity in empirical data is merely assumed without testing for it.

We must interpret the boundary conditions in order to understand how stability worked: it was not the gold standard, which is superfluous, but rather the serious threat of punishment combined with reward that stabilized the system. The band $b_1 < x < b_2$ represents the threat of government intervention. The process is asymptotically stationary iff. both b_1 and b_2 are finite (particle confined between two walls), so that $< x > = (b_1 + b_2)/2 = $ constant represents the gold value of the Dollar. The central bank would threaten to intervene to buy/sell Dollars if x would hit b_1 or b_2, so speculators would confidently buy Dollars if $\sigma < x < b_1$, e.g., where $\sigma^2 = < x^2 > =$ constant. Ergodicity of a stationary process guarantees that profitable fluctuations occur with average first passage time $\tau = \sigma^2/2\sigma_1{}^2$. *So it was the boundary condition, and not the gold standard itself, that provided stability.* Setting boundaries, establishing "threats, of punishment with teeth", made the difference. In those days the central bank had more gold than the speculators and so could the threat of intervention had to be taken seriously.

In the language of the Ultimatum Game [10], the boundary conditions/government regulations were threats of financial punishment. Those threats of punishment were effective ("had teeth") if the central banks had gold reserves large enough to beat the speculators. A related social analogy is the old saying "kids like boundaries". Also related is the old problem of 'The Tragedy of the Commons (Allmende)': with free farmers sharing a common meadow the tendency is for each farmer to add 'one more cow'. This is an example of an unregulated free market system. Adam Smith wrote earlier that moral restraint is required for a free market system to function. But mere moral restraint is inadequate, "regulations with teeth" are required for stability. Even during the depression, the gold reserves of the U.S., France, and Germany were more than adequate compared with the paper currency in circulation.

The system worked, speculators created stabilizing self-fulfilling prophecies before WWI, because governments (a) had adequate gold reserves and (b) saw their job as maintaining the stability of the currency, instead of guaranteeing high employment. WWI changed that. The rise of socialism and labor unions after WWI meant that social spending, *inflation via printing paper money, had to be given priority.*

Eichengreen asserts that post WWI FX markets were both volatile and (in our words) nonstationary: speculators, expecting any sign of devaluation of a currency to reflect the inability of a government to maintain the gold-value of the currency, bid the currency lower. That shift in agents' expectations is attributed in part to the fact that government spending on social programs was largely negligible before WWI, but had to be taken seriously by 1930. We have no evidence for volatility prior to 1987, but we do know empirically that by 1958 stock markets [11] were dynamically unstable/nonstationary [2]. 'Volatility' in a diffusive model like (2) requires an x-dependent diffusion coefficient [2]. The model of the 1958–1987 stock market, Osborne's Gaussian returns model, is unstable but not volatile.

The shift from stability to instability of the market noise distribution is an example of complexity: there was no way to foresee that development beforehand. The clearest mathematical idea of complexity to date is the absence of scaling laws (the absence of attractors and symbolic dynamics in a *deterministic* system) and instead the occurrence of surprises [12]. In a complex system, the future, given a precise initial condition, is not even statistically predictable.

Markets are complex even if we can describe the noise in a given era via simple stochastic dynamics [2, 3, 8]. We cannot describe the expected return R or the consensus price pc quantitatively in any financially useful way. Complexity enters financial markets in at least three ways: via the consensus price (the price where the returns density peaks), the expected return, and through unexpected shifts in the form of noise distribution. Eichengreen's thesis can be stated mathematically as the claim of a shift in the form of the noise distribution as a consequence of socialist/union pressures after WWI.

Summarizing, central bank intervention to devaluate was largely unnecessary before 1914, because if the Dollar fell to the gold export arbitrage point b_2 then speculators, believing the bank would intervene and maintain the gold value of the Dollar, bought the weak currency with the assurance of a later gain, preempting the necessity for the bank to devalue. *Derivatives (FX options) were unnecessary as hedge against foreign currency risk.*

2 FX Markets from WWI to WWII

Inflation of the money supply via credit is necessary but not sufficient for a boom or bubble [13]. Banks and the government saw the avoidance of inflation, not high employment, as their main job. Because of this particular social orientation, the stock market crash of 1929 caused a financial crisis: deflation occurred because there was no 'lender of the last resort' to provide liquidity. Margin trading in the 1920s, as in the 1990s, was a major source of running up stock prices, and when so the crash affected the lenders. Banks in that era were allowed to lend money for stock speculation, and many banks went bankrupt and closed after the Crash. Depositors withdrew money from remaining banks for fear of losing it, and so there was a liquidity draught. The depression/deflation followed from the lack of money in circulation: many people were unemployed, and those with money tended to hoard instead of spending it because *bank deposits were not insured in that time* (ask yourself why there has been no economic depression in the West since then, in spite of several stock market crashes). FDR was elected President in 1932 based on his promise/threat to abandon past social policies and institute a 'back to work' policy, therefore to restore liquidity. People with money knew what that meant, inflation via social spending, and so tried to convert their Dollars into gold.

FDR's famous Bank Holiday in 1933 was partly the consequence of a run on gold by speculators getting rid of the Dollar: they expected social spending and consequent inflation (gov't. spending was Keynes's recommendation for getting out of a slump). So, in 1935 FDR outlawed the ownership of gold by Americans, recalled all gold coins (excepting those held by coin collectors), and fixed the price of gold artificially at $35/oz., thereby guaranteeing that Americans could not depreciate the Dollar by buying gold. Fear of bank failures was widespread: my grandfather told me how his father dug up jars of gold coins from the back yard and traded them for paper money in 1935. In a further effort to control the Dollar, bank deposits were insured, and the Glass-Steagal Act was passed to keep banks out of the stock brokerage business. After a long time these acts eventually restored confidence in the banking system, but the hangover of fear from the Depression lasted into the 1950s. With Roosevelt's acts in place as law, the U.S. Gov't. could then spend freely on public works projects like TVA, which was and still is a very effective socialist project, and the WPA and get away with it, because the U.S. then had the largest gold reserves in the world. France and surprisingly, Germany, were second and third.[1] WWII brought new pressures into play.

3 The Post WWII Era

3.1 Era of 'Adjustable Pegged' FX Rates

After WWII Europe was broke. The U.S. ally Great Britain had to repay war loans while the enemies, Japan and Germany, were financially rewarded. Eastern Europe became a Soviet protectorate outside the realm of capitalism, while Germany and Japan under the Marshall Plan were rebuilt as capitalist bulwarks/showcases against

[1] A myth is that the Nazis came to power in 1933 because The Allies drained Germany via reparations. In fact, Germany successfully resisted reparations but had high unemployment for the same reason as did the US. Hitler's Finanz Minister Hjalmar Horace Greely Schacht got the ball rolling via Keynsian-style inflationary public spending. That was the era when the Autobahns began to be built, e.g. the wild German inflation of 1923 was a Berlin ploy to deflect France's demands for reparations payments.

the USSR and Communist China. The gold standard was replaced in the West by the Bretton Woods Agreement based on Keynesian ideas: (1) 'adjustable' pegged exchange rates, (2) controls were allowed and placed on international capital flows, and (3) the International Monetary Fund (IMF) was created to monitor economic policies within participating nations, and to extend credit to countries at risk with large trade imbalances.[2] Interest rates were capped as well. Controls were understood as necessary to avoid flight from a currency because the commitment to economic growth and/or full employment is not consistent with absence of inflation. Financial markets in that era were regulated, were not entirely free. And the expensive Arms Race with the USSR was on.

In 1957, Russia upped the Arms Race ante and frightened Americans by launching Sputnik. In October, 1957, my family and many others went out every clear night to watch Sputnik pass over. Our chief rocket scientist, Werner von Braun, complained that the U.S. bureaucracy had held him back, but finally launched a satellite from his site, Redstone Arsenal in Alabama, in January, 1958. One can certainly say that Sputnik inspired America to put a man on the moon. Von Braun had traveled from Penemünde to Reutte in Tirol (half an hour from where I partly now live) in 1945 to avoid capture by the Russians, and sent his brother to the Americans to give themselves up in order to avoid capture by the French. Without von Braun there would have been much less of a U.S. rocket program. When von Braun gave himself up to the American Army in Tirol in 1945[3] he humbly claimed only to have been following up on the rocket science invented by Robert Goddard. Today, one can visit the "V-2 Rocket Park" in White Sands, N.M., where an original V-2, the first ballistic missile, stands near the entrance. Quants' on Wall St. are also known as 'rocket scientists', and with good reason as we will explain below.

By 1959, the Bretton Woods (BW) exchange controls began to show signs of failure. The U.S. had pegged the Dollar artificially to internally nonliquid gold while inflating, and the rest of the West pegged currencies to the Dollar with the right to exchange Dollars for gold. I.e., the Dollar became the unit of international exchange. Financially and economically, to a first approximation, the Soviet Union and China did not exist except as military threats to stimulate arms spending. After WWII came the Marshall Plan and the Cold War (both the Arms Race and U.S bases in Europe were expensive), and by this mechanism U.S. gold reserves were in danger of being lost via massive postwar spending to rebuild Germany, Japan, and Western Europe generally. With the removal of some or the BW trade restrictions in 1959 Japan and Germany, instead of allowing more U.S. imports, restricted U.S. imports further. The result, taken all together, was that by 1959–1960 the order of magnitude of Dollars in Europe was on the order of magnitude of the Dollar value of gold stored in Ft. Knox, Ky. *We don't need a detailed dynamical model to understand that speculators expected a devaluation of the Dollar.* Consequently, Eisenhower (1958) prohibited Americans from owning gold in Europe, and Kennedy (1961)

[2] The traditional method of removing a serious trade imbalance was devaluation of the currency against gold.

[3] There is a tiny 'finger' of American invasion from Reutte to Ehrwald in North Tirol, which was later occupied by the French.

outlawed the ownership of gold coins by American collectors! This is the tale of an attempt by the US Gov't. to delay the inevitable: either transfer all the gold in Ft. Knox on demand to France, or else to devalue the Dollar. *OPEC was formed in the same era.*

Destabilizing speculative behavior is described by the Gaussian returns model that Osborne proposed empirically to describe stock returns (FX returns were not tested) in 1958 [11]

$$dx = (r - s_1{}^2/2)dt + s_1 dB(t) \tag{3}$$

subject to no boundary conditions. *Even with $r < 0$ this model allows no approach to statistical equilibrium* [2], is inherently nonstationary/unstable. Presumably, FX markets were like this since the end of WWI.

Modern credit, based on fractional reserve banking, is a partly uncontrolled form of money creation. The first credit card, the BankAmericard (later VISA) appeared ca. 1960, introduced by an Italian-American banker in Calif. presumably to help the local Italian community. By 1964 the Dollar was weak enough that silver coins were worth more than their face value, so the U.S. stopped minting them. The Vietnam War caused inflation, as do all wars.[4] So by 1971 France held enough Dollars quite seriously to demand payment in gold that would have cleaned out Ft. Knox. ARPANET appeared in that era, signaling that communication speed would increase in the future.

3.2 Emergence of Deregulation and Privatization as Policy

Under pressure from the fact that the order of magnitude of Dollars outside the U.S. was the order of magnitude of the U.S. gold supply, Nixon deregulated the Dollar in 1971 after first raising the official price of gold to $38. Significantly, Forex was also created in 1971, and the first ARPANET email program appeared then as well. The Chicago Board of Options Exchange (CBOE) was created in 1973, the year that the Black–Scholes equation and solution [14] were finally published after multiple rejections with editors and referees claiming, 'It's not economics'. The creation of derivatives/options/hedging exploded.

By 1973 gold had hit $800/oz., options trading was in full swing, and OPEC, correctly understanding the Dollar devaluation, raised the price of oil drastically. Several Middle Eastern countries also nationalized their oil fields. A VW Beetle that had cost $700–800 in 1968 cost $1,600 by 1974, and the gasoline price had doubled in Dollars as well, after first hitting $1.20/gal. All of that inflation was due to deregulating the Dollar from gold.

This was the beginning of deregulation in American politics. That program, globalization via privatization and deregulation, is now worldwide. Americans got rid

[4] No well-off population would ever choose war were the costs explained in advance, and were it understood that the war must eventually be paid for via either higher taxes, inflation, or generally by both.

of Dollars in favor of anything and everything (e.g., art, collectible coins) and ran up credit card bills with the expectation that tomorrow's Dollar would be worth much less than today's. In the 1980s Savings and Loans were 'deregulated' and went bankrupt, unable to compete with bond houses who split principle and interest into separate derivatives for speculation. Leveraged buyouts financed by junk bonds emerged and became the order of the day on Wall St. Monetization of nearly everything began in earnest, derivatives being a form of monetization and with credit/leverage providing the required pseudo-liquidity. FX transactions, on the order pf $\$10^8/\text{min}$. ca. 1981, decoupled from economic growth. With no certainty about currency at home or in international trade, physicists were hired on Wall St. to model derivatives (there was a big supply of unemployed academic physicists due to ten years of gov't. spending in response to Sputnik).

The inflation rate became so high ('stagflation', meaning inflation combined with unemployment) that the (Carter-appointed) U.S. Federal Reserve Bank Chairman Volcker let interest rates float to 13.5% in 1981, reducing inflation to 3.5% by 1983. Reagan-Thatcher-Friedman-omics began its (until recently) unchallenged heyday, with Mr. Reagan busting the U.S. budget worse than any President other than Mr. Bush.

Faster and cheaper communication encourages faster financial transactions. By 1985 Apple (and Commodore) computers and PCs were becoming common in offices and homes. Discount brokers appeared, also a form of finance market deregulation. In the 1970s and earlier, unless one had enough money to bring the buy/sell rates for stocks down to 3% or lower, one phoned one's stock broker and paid 6% coming and going, and orders were executed at a snail's pace. Limit orders were not very effective because of the slowness of execution, unless you were a big enough player to have a trader on the exchange floor. By 1999 one could place a small (say \$10K–30K) limit order with a discount broker on a Mac or PC and see it executed in a matter of seconds, if the order were placed close enough to the respective bid/ask prices (also shown in real time), paying \$20 to buy and with no charge to sell. Liquidity and speed of transaction have increased by many orders of magnitude.

In 1987 Peter Scholl-Latour [15] wrote about Afghanistan that if the Mujahedin were to get shoulder held rocket launchers, then the Russians would be finished. The Russians had the Mujahedin on the run in the Hindu Kush via their helicopters. I recall in 1988 seeing a photo of a CIA operative sitting, with big smile, on an Afghan hillside holding a rocket launcher. The Mujahedin systematically shot the Russian helicopters out of the sky, and in 1989 the Russians hightailed it out of Afghanistan. In 1990, bled by the Afghan War and the arms race with the U.S., the U.S.S.R. collapsed.[5] Consequently, the lid blew off two pots that had been boiling for a long

---

[5] In 1989 the U.S. and G.B. gave should held rocket launchers to the Mujahedin. Russian helicopters could no longer successfully fight in the Hindu-Kush. A year later the Russians left Afghanistan, defeated. The Taliban inherited the should-held rocket launchers, and they still work. No in power one understood in that time that the centralized USSR was far less a threat to the West than would be delocalized Islamic Fundamentalism. We live in an era of religious fundamentalism, Christian, Islamic, Hindu, and Jewish.

time: it was expected that markets would then become globalized, but western leaders did not foresee that Islamic Fundamentalism (fomenting in Pakistan and Egypt since the 1930s) would explode onto the world as well, with terrorism plotted by medieval mentalities financed by oil money, and using modern technology to fight the globalizing influences of capitalist tech-culture. The Taliban later used the CIA-(with the help of the ISS) supplied rocket launchers. Capitalists did not understand or foresee that delocalized Islamic Fundamentalist networks would be a threat to the world replacing the threat of centralized communism.

In 1994 an extremist free market Congress took power in the U.S. (they believed that privatization/deregulation would be the best thing since white bread) and the stock bubble grew correspondingly: the DJA quadrupled from 1994–2000 (from 1987–1994 it had doubled, and had doubled earlier from 1973–1987). The signs of the bubble were everywhere: working people quit their jobs to 'momentum trade' dot.com stocks on margin. *The bubble was popped in exactly the same way as in 1929*: the Fed, seeing the bubble, tried to deflate it slowly via a systematic sequence of many small quarterly interest rate increases. Econophysicsts and financial engineers have no idea how to model this effect correctly. The difference with 1929 is that, with lenders of the last resort in place to avoid a crash, the air never completely came out. Stocks still sell at very high price/earnings ratios because there are so few other places to park all the money in circulation. But as we've pointed out elsewhere, ideas of 'overvalued' and 'undervalued' are effectively subjective [4].

In 1998 the world finance market nearly crashed catastrophically [16]. Following two ideas, the Modigilani–Miller 'theorem' and the expectation that market equilibrium will prevail after large deviations, the hedge fund Long Term Capital Management (LTCM) had achieved a debt to equity ratio 'approaching infinity', with leveraging supplied by nearly every major bank in the world. The fund was run by bond trader John Merriweather and Economics Nobel Prize winners Scholes and Merton, of Black–Scholes–Merton fame (Black died in 1993 before the Nobel Prize for the Black–Scholes equation was awarded). Ignoring that markets are nonstationary and that the liquidity bath is necessary for the application of Brownian models in the first place, that hedge fund literally 'became the market' in Russian bonds (with trading rules based on the Black–Scholes model), so that when they wanted to sell they suffered the Gamblers' Ruin. Modigiliani and Miller had considered only small changes in returns, where there is no need to worry about the Gamblers' Ruin.

Today, instead of gold, and central bank 'sticks' and 'carrots' for speculators, we have the IMF and other nondemocratic, supra-governmental agencies like the World Bank and the WTO that try to play a regulatory role in the world economy (the EU is not a confederated democracy like Switzerland, but is top-down bureaucracy, or administration (*Verwaltung*), with rules made in Brussels handed down by edict). A currency remains strong when an economically strong enough country pays higher interest rates, which is a deflationary mechanism. A weak currency like the Bush Dollar, with trade and budget deficits running through the ceiling since 2001, can finance its debt only through attracting foreign money via high enough interest rates. A small country like Argentina cannot do that: the U.S gets away with it only because speculators do not believe that the U.S. Gov't. will allow the Dollar to collapse, which reminds us vaguely of the way that the gold standard worked before WWI. I.e., the whole edifice is built on belief.

There is in any case not enough gold in the world even to think about financing modern consumption and economic growth. The nr. of Dollar increased about 55% from 1945–1965, and by about 2000% from 1971–2001. Inflation via fast, cheap, credit is necessary for what we call 'modern prosperity'.

Money is literally created by the tap of a computer key when you use your credit card. Governments in our era use interest rates to try and keep this in check. For reference, when I first went to Germany in 1985, one could not buy a car on credit, and credit cards were useless in restaurants and gasoline stations, could only be used in tourist businesses in big cities.

Mr. Bush systematically eliminated the 2000 budget surplus and ran up a deficit via tax rebates, lowered taxes, and increased spending, on the basis of the extremist (neo-con) philosophy that a government in the red has a good excuse to eliminate all social programs. The neo-con program for America includes ending the financing of public schools. In the eyes of the free market extremists, everything should be privatized. Even the U.S. Army has been to a high degree privatized, with Halliburton paid to supply food and other services, previously taken care of by Army logistics, in Iraq and Afghanistan.

For the U.S., the traditional method of remedying a trade imbalance (devaluation of the currency) no longer works. With the Bush deficit made worse by war spending, the Dollar fell from $0.87/Euro in 2000 to $1.33/Euro today, but China, beyond control by the West, pegs the Yuan to the Dollar, guaranteeing that cheap production in China will always win. The West exports manufacturing to China, increasing unemployment and simultaneously increasing inflation via increasing oil prices due to a weak Dollar.[6] At the same time, the U.S. must pay high enough interest rates to attract foreign capital (via sale of Treasury Bills and Bonds) to finance the enormous deficit. China had enough dollars to buy Chevron in 2005, but the free market congress nixed the deal. A central bank's method of trying to control the money supply is simple: selling bonds and bills reduces the money supply (banks are required to participate), while buying them back increases the money supply. With fractional reserve banking, necessarily required for the inflation via credit for economic growth and consequent consumption, a bank need keep on hand only a certain fraction of what it lends (again, without full backing by gold, credit is money creation). The story with China now is similar to the story with Europe in the 1950s: the U.S. sacrifices it's currency (and now jobs as well) for strategic reasons. A trade tariff would be the only alternative.

The history of globalizing capital can therefore be seen systematically as the history of the increase in liquidity and deregulation internationally. It was, in fact, only a few years ago that the Glass-Steagal act was eliminated by the believers

---

[6] According to Milton Friedman, if someone can make it cheaper then let him make it. That extremist free market position makes an invalid assumption about 'liquidity', namely, that via inventiveness enough new jobs will be found to replace those lost. But economists have not worried enough about empirical and theoretical inconsistencies in their assumptions.

in completely unregulated free markets who dominate the U.S. Congress, allowing banks to go back into the stock market business. The entire program in deregulation is based on an uncritical belief in a nonexistent stability of unregulated markets, in the face of empirical evidence that unregulated markets are unstable.

3.3 Emergence of Finance Theory and Econophysics

With the above history of FX markets in mind, I turn now to the parallel history of finance theory and econophysics.

Around 1930, Mr. Cowles, who had enough money to finance research in a serious way, asked the question, 'Can stock prices be predicted?' He then funded the journal Econometrica and the Cowles Foundation. But in a sequence of developments described by Mirowski in "Machine Dreams" [17], neo-classical economic theory, stuck in irrelevant equilibrium mathematics, became dominant while econometrics got stuck with empirically irrelevant stationarity assumptions, so that economics research until this day has nothing to offer us. Consequently, finance market modeling developed completely independently of (and in opposition to) existing economic theory. There is still no connection between the two, the former being empirically based and important for banks and brokerage houses, the latter being completely wrong but necessary for academic advancement in the economics priesthood, and (unfortunately for all of us) provides the mathematized ideology for governments and agencies who believe that privatization and deregulation of nearly everything will lead to a better world (World Bank, I.M.F., W.T.O., E.U., U.S. Treasury and U.S. Federal Reserve Bank, etc.).

The first falsifiable model of relationship applicable to a finance market was due to Markowitz (ca. 1950) and was simplified by Sharpe (ca. 1960). Given the market distribution of log returns of a financial object (stock, bond, or foreign exchange), $x(t) = \ln p(t)/p_c$ where p_c is a reference price, assume that the distribution has a finite second moment. The variance is taken to measure uncertainty. Form a hypothetical portfolio of n paper assets and then minimize the variance for a given expected return. The predicted relationship between expected return and variance is called the Capital Asset Pricing model, or CAPM. CAPM has been falsified [2]. CAPM motivated Fischer Black to try to derive an option pricing model, but CAPM does not reproduce the Black–Scholes pde, the underlying psychological assumptions about traders are different in CAPM than in the Blsck–Scholes delta hedge.

In 1958 Modigliani and Miller [18] published their Nobel Prize winning paper announcing two results: (i) that the debt/equity ratio B/S is irrelevant in the evaluation of a firm, and (ii) that paying or not paying dividends has no effect on stock prices. The latter is correct, the former claim is not. The derivation of the notion that the debt to equity ratio B/S is irrelevant makes a small returns approximation, and makes an illegal liquidity assumption *while ignoring liquidity altogether* (noise is liquidity, and returns were treated as behaving deterministically favorably) [2].

The next step, made by Osborne, was far more fruitful. Osborne can and should be regarded as the first econophysicist. He plotted histograms of stock prices listed daily in the Wall St. Journal and concluded that they seemed to be approximately log normally distributed [11], meaning that returns density f(x,t) is Gaussian (1958). This model was used by Black and Scholes to price options, and was falsified only after the stock market crash of October 1987. *Osborne's model describes a nonstationary, i.e. dynamically unstable market: the returns variance increases with time t, meaning that statistical equilibrium is not approached.*

In the same era, Mandelbrot analyzed historic cotton prices [19] and concluded (wrongly, we think) that the variance is ill-defined, so he advised setting the variance equal to infinity. He correspondingly suggested Levy distributions to describe the statistics. Levy distributions have infinite variance and fat tails, $f(x,t) \approx x^{-\mu}$ if $x \gg 1: \ 2 < \mu < 3$. We know that FX markets have finite variance that varies in a complicated but systematically repeated way during each day with time [3]. Levy aside, our quadratic diffusive model [20] generates fat tails with $2 < \mu < \infty$, with finite variance for $3 < \mu < \infty$. A definitive statement about fat tails will be given below.

Mandelbrot introduced two more ideas that are more for finance: fractional Brownian motion (fBm) [21], which requires stationary (and therefore correlated) increments, and the idea of the efficient market hypothesis (EMH) as a martingale in the returns variable [22], which required uncorrelated and therefore generally nonstationary increments (this is explained in detail below). The EMH is described by martingales, and fBm is not a martingale; fBm violates the EMH.

Fractional Brownian motion (fBm) has *two* defining features [21, 23]: (i) stationary increments $x(t,T) = x(t+T) - x(t) = x(T)$ and, with $x(0) = 0$, (ii) variance scaling, $< x^2(t) >= t^{2H} < x^2(1) >$. Although nearly every paper you'll read in the literature ignores (i) and emphasizes (ii), stationary increments combined with variance nonlinear in t imply long time increment autocorrelations, $< x(t,T)x(t,-T) > \neq 0$, and scaling (ii) is unimportant. Scaling, taken alone, does not imply any correlations at all, nor does scaling define a stochastic process. Markov processes have vanishing increment autocorrelations and can scale with Hurst exponent H, demonstrating that Hurst exponent scaling does *not* imply nontrivial autocorrelations. This is explained in detail below.

The idea of a martingale is subtle but physically interesting. Martingales, by definition, obey a conditional average $< x(t) >_c = \int p_2(x,t:y,s)x dx = y$, where p_2 is the transition probability density, or conditional probability density. Martingales 'look' Markovian because they have vanishing increment autocorrelations (generally with nonstationary increments) but can have finite memory (fBm memory is infinite). A Markovian market would be impossible to beat. Real finance markets are very hard to beat. The martingale property reflects this idea. A market with noise described by fBM is in principle beatable because one could trade on the increment autocorrelations.

In 1973 Black, Scholes [14] and Merton quasi-independently published the first falsifiable model of option pricing (derivatives pricing) based on the delta hedge, or risk neutral condition. The model assumes Markovian (i.e., diffusive) Gaussian returns and is based on two empirically measurable parameters. The model was falsified after the 1987 Crash. Whether it held accurately before 1987 we don't know, the pre-1990 data are not good enough to test the model. We do know that post 1987 finance markets are volatile. The delta hedge can be shown quite generally [24] to represent a fair game condition (martingale).

In 1977 Osborne [25] took another giant step for our understanding of economics and finance: he falsified the neo-classical supply–demand curves, predictions that based on the nonfalsifiable and irrelevant idea of utility maximization, and explained that supply–demand curves are instead step functions. In his book, he presents limit book orders as an example and also explains how limit book orders and market making work.

An extremely important observation made by Black, who had studied physics at Harvard before going into applied math and then finance (Black rose to Partner status at Goldman-Sachs). Black introduced the term 'noise traders' [26], reflecting the idea first made explicit in Osborne's lognormal pricing model that the market is essentially noise. Noise traders provide the liquidity, buying and selling often and apparently randomly. I.e., they provide the liquidity necessary for 'Brownian' market models. I pointed out in 2003 that liquidity (the money bath) is the market analog of the heat bath of a Brownian particle, although equilibrium statistical mechanical ideas cannot be applied to the money bath [2]. Liquidity is described by the entropy of the returns distribution.

Sputnik caused the U.S. government to fund physics, math and engineering via NDEA loans and research grants after 1958. I obtained the former every year as undergraduate, and went through graduate school on research fellowships. At Yale in 1965–1967 we even had full research fellowship support for the first year and a half before we had a research advisor! Sputnik paid lavishly for my graduate education. By 1970 there was an oversupply of physicists, too many for academic positions to absorb. In 1981 I first heard at a Los Alamos nonlinear dynamics meeting that one of Kadanoff's former grad students in nonlinear dynamics had gone to work for a Chicago trading house. Fired by Black–Scholes, the deregulation of option trading, the instability of markets, and the oversupply of physicists, this was the beginning of physicists in finance. In June, 2000, when I gave a talk at the modellers at Enron in Houston (televised to Enron in London), 26 of those modellers were from physics and engineering, two were from math (none were from economics or finance). The section heads were former experimental physicists! That tale of that era is told extremely well by Emanuel Derman [27], a collaborator of Black, who was one of the first physicists hired on Wall St.

In 1990 my close colleague Gemunu Gunaratne spent a year working for TradeLink Corp., a Chicago trading house, before coming to the University of Houston. There, he ignored advice from academic finance theorists (who insisted that the Black–Scholes equation explains option pricing) and instead talked with the traders (who explained that it doesn't). Consequently, he (i) discovered that FX data are exponential in log returns $x = \ln p(t)/p_c$, and then (ii) priced options in agreement with traders' prices. This became the basis for our further work when, in 2003, I discovered the dynamics of the exponential model. Gemunu corrected a mistake in my first attempt, and then I took it a step further afterward. Finally we discovered the Markov scaling solutions that allow one to calculate 1-point densities (but not transition densities) in closed form. Gemunu's 1990 paper was never submitted because he felt that he should first work out the dynamics of the model. I first saw that paper in 1996 when I had no idea whatsoever about options or Black–Scholes, and Gemunu's groundbreaking work forms the first part of Ch. 6 in my book [2]. He suggested that I publish it.

In 1995 Gene Stanley coined the phrase 'econophysics' at a meeting in India. At about the same time, Yi-Cheng Zhang and his group introduced the Econophysics web page, www.unifr.ch/econophysics. Simon Capelin at Cambridge University Press led the charge by publishing Stanley's and other econophysics books. His program there is still strong. Sadly, econophysics, like nonlinear dynamics, was never recognized by the APS as a part of physics. American physics prefers (nonfalsifiable) string theory to falsifiable approaches to social problems. My university offers one of the only Econophysics Ph.D. programs in the world.

Essential references on the history of financial markets are Bernstein [28], Lewis [29], Dunbar [16], and Derman [27].

4 The Mathematics and Markets of Globalization via Deregulation and Privatization

4.1 Expectations of Scaling

Many statistical physicists went into economics and finance in search of universality and scaling. There was never any good reason to expect markets to produce universal scaling exponents. Many papers have been published on both fat tail and Hurst exponents and they are nearly all wrong. Only the 1997 finance analysis of Zhang et al. [30] discovered a correct idea, and no one picked up on it until we realized early in 2006 that we were contradicting ourselves in our own data analysis (we'd discovered the importance of nonstationary increments theoretically, but then ignored that idea completely in our data analysis!). We 'discovered' the Zhang paper after we'd reproduced their main result, plus one [3, 31]. Because no one understood that paper, and also because nearly everyone incorrectly assumed that $H \neq 1/2$ implies long time autocorrelations (even though Mandelbrot and van Ness had pointed out the necessity of stationary increments), the literature is replete with wrong theoretical claims about scaling, and with correspondingly wrong data analyses.

I'll now summarize our latest empirical results [3]. The mathematics described in the next three sections is not restricted to finance markets but applies generally to the analysis of any random time series x(t) whether in physics, finance, biology, chemistry, or elsewhere. *The main point is that nearly all previous time series analyses should be redone: both Hurst and fat tail scaling are generated systematically as artifacts of a wrong method of data analysis.*

In 2000 the Euro cost \$ = 0.87. In 2006 the Euro cost \$ = 1.33. 'Value' p_c has shifted by about 50% but the noise distribution did not change. We find instability, the variance does not approach a constant but varies in a complicated way with time. The variance doesn't scale over the time interval of a day or a week, the returns distribution is exponential with no fat tails. The returns increments for nonoverlapping time intervals vanish for trading times greater than 10 min., and the increments are nonstationary. In fact, we observe a martingale, a diffusive stochastic process. Our results are based on a six year time series of Euro/Dollar prices where, because the increments are nonstationary we must view each day as a 'rerun' of the same 'experiment'. In fact, the weekly data indicate that that is a good assumption, that the same dynamics is repeated every day of the week [3]. That there are no fat tails, and no Hurst exponent scaling either, disagrees with most existing financial data analyses. The reason for this is the assumption in other analyses of stationary increments. The following sections explain scaling, fat tails, and stationary vs. nonstationary increments.

4.2 *Scaling with Hurst and Fat Tail Exponents H and* μ

A stochastic process x(t) scales with Hurst exponent H if [23, 32]

$$x(t) = t^H x(1) \tag{4}$$

$0 < H < 1$, where by equality we mean equality 'in distribution'. We see that $x(0) = 0$ occurs deterministically, regardless of all the different possible past histories of x(t). Hurst exponents H occur both in completely uncorrelated processes (Markov processes) and in strongly correlated processes like fBm.

For a specific stochastic process x(t), the 1-point density $f_1(x,t)$ of can be used to calculate the average of any dynamical variable A(x,t):

$$\langle A(x,t) \rangle = \int A(x,t) f_1(x,t) dx \tag{5}$$

From (4), the moments of x must obey

$$\langle x^n(t) \rangle = t^{nH} \langle x^n(1) \rangle = c_n t^{nH} \tag{6}$$

Combining this with

$$\langle x^n(t) \rangle = \int x^n f_1(x,t) dx \tag{7}$$

we obtain [23, 24]

$$f_1(x,t) = t^{-H} F(u) \tag{8}$$

where the scaling variable is $u = x/t^H$. In particular, with no average drift, so that we can choose $< x(t) >= x(t_o) = 0$, the variance is simply

$$\sigma^2 = \langle x^2(t) \rangle = \langle x^2(1) \rangle t^{2H} \tag{9}$$

None of this has the slightest thing to do with long time autocorrelations: scaling is neither necessary nor sufficient for long time autocorrelations like those of fBm. Scaling is essentially irrelevant for the question of EMH violating correlations that one might trade on.

Consider a diffusive process where the drift vanishes (martingale condition). The Ito stochastic differential equation (sde) generating the process is

$$dx = \sqrt{D(x,t)}dB \tag{10}$$

where B(t) is the Wiener process, $< dB^2(t) >= dB^2(t) = dt, < dB(t)dB(s) >= 0$ if $s \neq t$. Then with $x(0) = 0$ (always a necessary condition for scaling) we have

$$\sigma^2 = \int_0^t ds \int_{-\infty}^{\infty} dx g(x,s:0,0)D(x,s) \tag{11}$$

where g is the Green function/transition density for (10), so that, by (8) and (9), and with $g(x,s:0,0) = f_1(x,s)$, scaling implies that

$$D(x,t) = t^{2H-1}D(u), u = x/t^H \tag{12}$$

A class of Markov scaling solutions with scaling including Hurst exponent scaling [8, 23] is given as follows: let

$$f_1(x,t) = \sigma_1^{-1}(t)F(u) \tag{13}$$

with initial condition $f_1(x,0) = \delta(x)$, where $u = x/\sigma_1(t)$, with variance

$$\sigma^2(t) = \langle x^2(t) \rangle = \sigma_1^2(t) \langle x^2(1) \rangle \tag{14}$$

Then with the diffusion coefficient scaling as

$$D(x,t) = (d\sigma_1^2/dt)\bar{D}(u) \tag{15}$$

where $d\sigma_1/dt > 0$ is required, $f_1(x,t)$ satisfies the Fokker-Planck pde

$$\frac{\partial f_1}{\partial t} = \frac{1}{2}\frac{\partial^2(Df_1)}{\partial x^2} \tag{16}$$

and yields the scale invariant part of the solution

$$F(u) = \frac{C}{\bar{D}(u)} e^{-\int udu/\bar{D}(u)} \tag{17}$$

It's easy to prove that Hurst exponent scaling $\sigma_1(t) = t^H$, $0 < H < 1$, is the only possibility. A piecewise constant drift R(t) can be included in our result by replacing x by $x - \int R(s)ds$ in u [20, 33].

By fat tails we mean that $f(x,t) \approx x^{-m}$ if $x >> 1$. Fat tails are generated by diffusive processes scaling with H via $D(u) = 1 + eu^2$, $u = x/t^H$, where m and H are completely independent parameters [20]. The scale invariant part of the 1-point density is then student-t-like. Were fat tails to occur empirically in FX data then option prices would blow up [24].

One cannot identify a stochastic process by a scaling exponent, and more generally not by a 1-point density [23] or a pde for a 1-point density [34]. E.g., *1-point densities* for drift-free Markov processes, fBm and martingales all obey the same pde, but the corresponding 2-point conditional densities do not. The transition density for fBm obeys no pde at all. Much confusion has been propagated throughout the literature under the misconception that a 1-point density of pde for a 1-point density defines a stochastic process, or even a class of stochastic processes. Essentially, a 1-point density tells us nothing whatsoever about the dynamics of the time series x(t) under observation [34, 35].

4.3 Stationary vs. Nonstationary Increments

In the analysis that follows, we assume a drift-free nonstationary process x(t) with the initial condition $x(0) = 0$, so that the variance is given by $\sigma^2 =< x^2(t) >= \int x^2 f_1(x,t)$. By the increments of the process we mean $\Delta x(t,T) = x(t+T) - x(t)$ and $\Delta x(t,-T) = x(t) - x(t-T)$. In what follows we will use the simpler notation $x(t,T) = \Delta x(t,T)$. The question is: when is x(t,T) a "good" variable (when can x(t,T) be used as a coordinate?).

Stationary increments [8, 21, 23] are defined by

$$x(t+T) - x(t) = x(T) \tag{18}$$

'in distribution', and by nonstationary increments [5, 8, 23] we mean that

$$x(t+T) - x(t) \neq x(T) \tag{19}$$

in distribution. When (18) holds, then $x(t,T) = x(T)$ is a good variable, and given the density of 'positions' $f_1(x,t)$, we also know the density $f_1(x(T),T) = f_1(x(t+T) - x(t),T)$ of increments independently of the starting time t. Whenever the increments are nonstationary then any analysis of the increments inherently requires the 2-point density, $f_2(x(t+T),t+T;x(t),t)$. From the standpoint of theory there exists no 1-point density of increments f(x;T),T) depending on T alone, independent of t, and spurious 1-point histograms of increments are typically constructed empirically by assuming that the converse is possible. Next, we place an important restriction on the class of stochastic processes under consideration.

According to Mandelbrot, so-called 'efficient market' has no memory that can be *easily* exploited in trading [22]. Beginning with that idea we can assert the necessary but not sufficient condition, the absence of increment autocorrelations,

$$\langle (x(t_1) - x(t_1 - T_1))(x(t_2 + T_2) - x(t_2)) \rangle = 0 \tag{20}$$

when there is no time interval overlap, $t_1 < t_2$ and $T_1,\ T_2 > 0$. This is a much weaker condition and far more interesting than asserting that the increments are statistically independent. We will see that this condition leaves the question of the dynamics of x(t) open, except to rule out processes with increment autocorrelations, specifically stationary increment processes like fBm.

Consider a stochastic process x(t) where the increments are uncorrelated. From this condition we easily obtain the autocorrelation function for positions (returns), sometimes called 'serial autocorrelations'. If $t > s$ then

$$\langle x(t)x(s)\rangle = \langle (x(t)-x(s))x(s)\rangle + \langle x^2(s)\rangle = \langle x^2(s)\rangle > 0 \tag{21}$$

since with $x(t_o) = 0$ $x(s) - x(t_o) = x(s)$, so that $< x(s)x(t) > = < x^2(s) >$ is simply the variance in x. Given a history $(x(t),\ldots,x(s),\ldots,x(0))$, or $(x(t_n),\ldots x(t_k),\ldots,x(t_1))$, (21) reflects a martingale property. If f_n denotes the n-point density and $p_2(x_n,t_n : x_{n-1},t_{n-1};x_{n-2},t_{n-2};\ldots;x_1,t_1)$ denotes the conditional density (transition density) depending on history $(x_{n-2},t_{n-2};\ldots;x_1,t_1)$, then

$$\begin{aligned}\langle x(t_n)x(t_k)\rangle &= \int dx_n\ldots dx_1 x_n x_k p_n(x_n,t_n\,|x_n,t_n,\ldots,x_n,t_n,\ldots)\, p_{n-1}(\ldots)\ldots p_{k+1}(\ldots)\\ &\quad\times f_k(\ldots) = \int x_k{}^2 f_k(x_k,t_k;\ldots;x_1,t_1)dx_k\ldots dx_1\\ &= \int x^2 f_1(x,t)dx = \langle x_k^2(t_k)\rangle\end{aligned} \tag{22}$$

where

$$\int x_m dx_m p_m(x_m,t_m\,|x_{m-1},t_{m-1}\,;\ldots;x_1,t_1) = x_{m-1} \tag{23}$$

Every martingale generates uncorrelated increments and conversely, and so for a Martingale $< x(t)x(s) > = < x^2(s) >$ if $s < t$.[7]

What follows next is crucial for avoiding mistakes in data analysis.

Combining

$$\langle (x(t+T)-x(t))^2\rangle = +\langle (x^2(t+T)\rangle + \langle x^2(t)\rangle - 2\langle x(t+T)x(t)\rangle \tag{24}$$

with (18), we get

$$\langle (x(t+T)-x(t))^2\rangle = \langle x^2(t+T)\rangle - \langle x^2(t)\rangle \tag{25}$$

which depends on both t and T, excepting the case where $< x^2(t) >$ is linear in t. Uncorrelated increments are generally nonstationary. Therefore, martingales generate uncorrelated, typically nonstationary increments. So, at the level of pair correlations

[7] Note that (22,23) hold for time translationally invariant martingales, where $p_2(x,t:y,s) = p_2(x,t-s:y,0)$. One can easily check this for a drift-free Gaussian Markov process. I.e., time translational invariance does not imply that $< x(t)x(s) >$ is a function of t-s alone. Time translational invariance of p_n, $n \geq 2$, does not imply that a statistical equilibrium density $f_1(x)$ exists *and is approached asymptotically* by $f_1(x,t)$ [9]. I.e., a time translationally invariant martingale on $[-\infty,\infty]$ cannot yield a stationary process, cannot lead to statistical equilibrium.

a martingale with memory cannot be distinguished empirically from a drift-free Markov process. To see the memory in a martingale one must study at the very least the 3-point correlations. The increments of a martingale may be stationary iff. the variance is linear in t (we restrict ourselves to the consideration of processes with finite variance), but even if the variance is linear in t there is no guarantee a priori that the increments are stationary.

A martingale x(t) has no drift, and conditioned on the return $x(t_o)$ yields $< x(t) >_{cond} = x(t_o)$. That is, x(t) not only has no trend but the conditional average is in addition 'stuck' at the last observed point in a time series,

$$\int x_n p_n(x_n, t_n | x_{n-1}, t_{n-1}; \ldots; x_1, t_1) dx_n = x_{n-1} \tag{26}$$

Since x(t) represents the return or 'gain', one further event in the sequence produces no expected gain.

4.4 Analyzing Time Series with Nonstationary Increments

Earlier empirical FX market data analyses disagree on details but nearly all report that the returns distribution is non-Gaussian, scales with Hurst exponent H $f(x,t) = t^{-H}F(u)$, and has fat tails for large returns $f(x,t) \approx x^{-m}$ for $x >> 1$ [2, 36, 37, 38, 39]. Here's our main point: *all but one* [30] *of the earlier data analyses are wrong because they use of a technique called 'sliding windows'* [3, 40].

This is what's meant by the sliding window method [30, 40, 41]. Start with a time series x(t). To obtain histograms for the variable x at different times T, one begins with the time series at time t and forms a window at later time $t+T$. One studies the increment $x(t,T) = x(t+T) - x(t)$ by sliding the window along the entire length of the time series, increasing t one unit at a time while holding T fixed. In finance, this is the same as assuming that $x(t,T) = \ln(p(t+T)/p(t))$ is a "good" variable, that x(t,T) is independent of t. This assumption that is made in nearly every paper you can find in the literature, *but it is wrong if the time series has nonstationary increments. And FX series do have nonstationary increments.* The method is correct for stationary increment processes like fBm, processes with long time increment autocorrelations.

For a long time series, one of at least $t_{max} \approx$ several years in length, the sliding window method is expected to produce good statistics because it picks up a lot of data points. Note that it's necessary to assume that the histograms generated from varying t in the *increments* x(t,T) yield f(x,T), the density of *returns* x(T), independent of the empirical time average over t. *This assumption is correct iff. The increments are stationary, iff.* $x(t,T) = x(T)$, *otherwise the assumption is false.* To repeat, the papers using sliding windows assume a method that implicitly requires the long time correlations of fBM, whereas (in contradiction) FX data at least approximately obey the EMH.

The reason why an increment cannot serve as a 'good' coordinate for a time series with nonstationary increments is that it depends on the staring time t: let $z = x(t;T)$. Then

$$f(z,t,t+T) = \int f_2(y,t+T;x,t)\delta(z-y+x)dxdy \tag{27}$$

is not independent of t, although attempts to construct this quantity as histograms in data analysis via 'sliding windows' implicitly presume t-independence of f. If the increments are stationary then $z = y - x = x(T)$ and we obtain a well defined density f(z,T). When the increments are nonstationary then f depends on t and (27) can at best be understood as a failed attempt to coarse grain.

In all that follows both the simulation and market data have been detrended: we work only with Martingale processes $< x(t) > = x(t_o)$.

To illustrate how spurious results are generated by using sliding windows in data analysis, we apply that method to a time series with *uncorrelated nonstationary increments*, one with no fat tails and with a Hurst exponent $H = 0.35$, namely, a time series generated by the exponential model (16) with $H = 0.35$ (Fig. 1a). Figure 1a

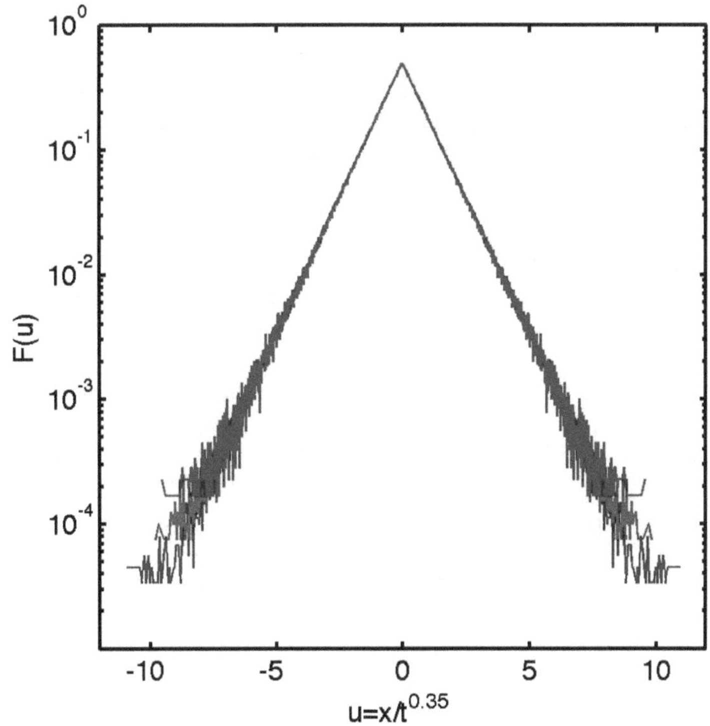

Fig. 1a The scaling function F(u) is calculated from a simulated time series generated via the exponential model, $D(u) = 1 + \text{abs}(u)$ with $H = 0.35$. 5,000,000 independent runs of the exponential stochastic process were used

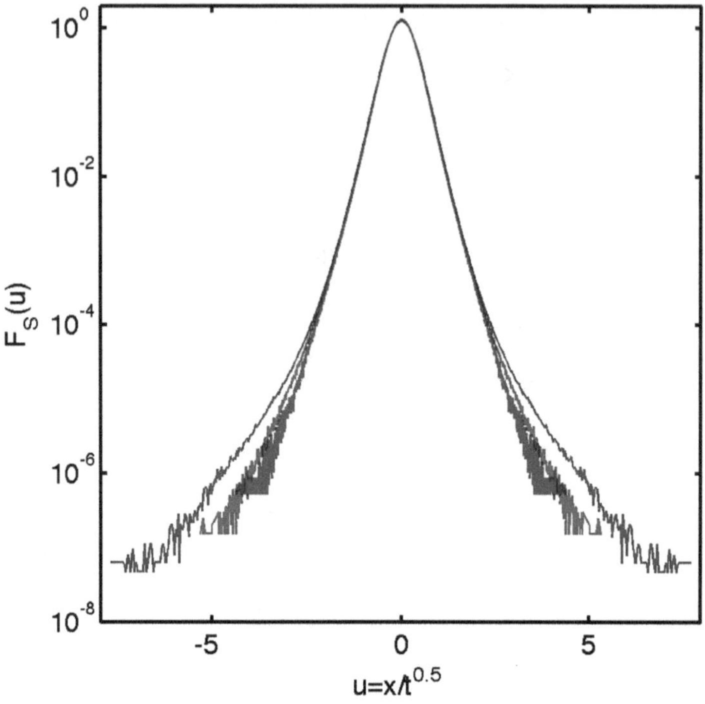

Fig. 1b The 'sliding window scaling function' $F_S(u_s)$, $u_s = x_s(T)/T^{H_s}$ was calculated for the same simulated data. Note that F_S has fat tails whereas F does not, and that $H_S = 1/2$ appears contradicting the fact that $H = 0.35$ was used to generate the time series. That is, sliding windows produce two significantly spurious results

was generated by taking 5,000,000 independent runs of the stochastic process, each starting from $x(0) = 0$ for $T = 10$, 100, and 1000. Here, then, $x(0,T) = x(T)$ satisfies the sde for the exponential process. The sliding window result is shown as Fig. 1b.

In this case, the sliding windows appear to yield a density $F_s(u_s)$, $u_s = x_s(T)/T^{H_s}$, from an empirical average over t that one cannot formulate theoretically, because for a nonstationary process there is no ergodic theorem. In fact, in order to try to derive a density of increments $x(t,T) = x(t+T) - x(t)$, we would have to start with the two-point density $f_2(x(t+T), t+T; x(t), t)$ and try to get rid of t, but there's no procedure for this, and there is good reason to expect that a 1-point density of increments does not exist. Note that F_s is badly behaved for large u, but as we have said the assumption that a distribution $F_s(u_s)$ $u_s = x(T)/T^{H_s}$ exists mathematically may be faulty. Summarizing the main points, not only are fat tails generated artificially, but we get a spurious Hurst exponent $H_S = 1/2$ as well, all based on a density F_s that may not be well defined mathematically. *This is the method that has been used in nearly all existing finance data analyses.* Furthermore, we can explain theoretically from simple calculations exactly how the spurious Hurst exponent and fat tails arise [41].

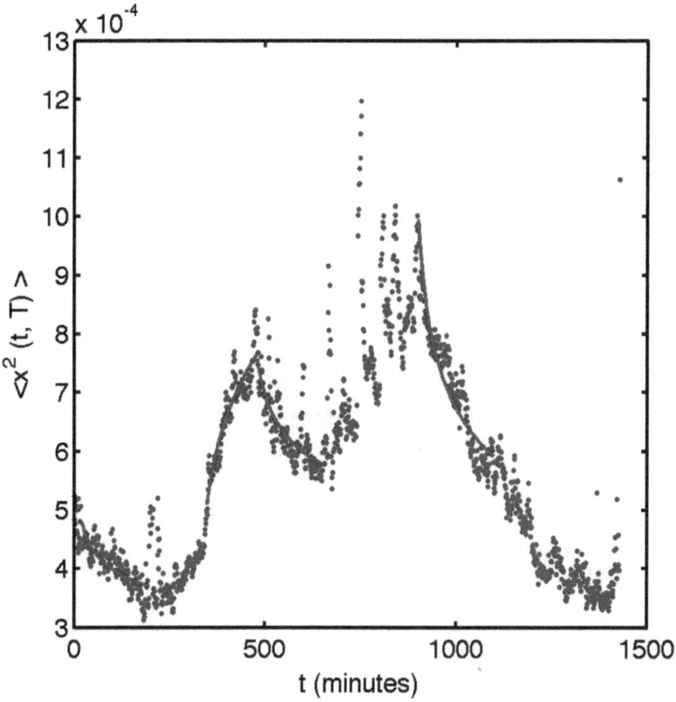

Fig. 2a The root mean square fluctuation $< x^2(t,T) >^{1/2}$ of the daily Euro–Dollar exchange rate is plotted against time of day t, with $T = 10$ min. to insure that autocorrelations in increments have died out (Fig. 3)

Next, we describe our study of a six year time series of Euro-Dollar exchange rates from Olsen & Associates [3]. The increments $x(t,T) = x(t+T) - x(t) = \mathrm{lm}(p(t+T)/p(t))$ are nonstationary, as is shown by the root mean square fluctuation in increments plotted against t in Fig. 2a, where $T = 10$ min. to insure that there are no autocorrelations in increments (Fig. 3).

First, note that the increments depend on t and therefore are nonstationary. Second, note that the returns data do not scale with a Hurst exponent H or even with several Hurst exponents over the entire day. Figure 2b shows that the same stochastic process is repeated on different days of the week, so that we can assume a single, definite intraday stochastic process x(t). In intraday returns Fig. 1a, scaling is observed at best within four disjoint time intervals during the day, and even then with four different Hurst exponents ($H < 1/2$ in three of the intervals, $H > 1/2$ in the other). That is, the intraday stochastic process x(t) generally does not scale and is instead complicated.

Within the three windows where a data collapse $F(u) = t^H f(x,t)$ is weakly indicated, we see that the scaling function F(u) has no fat tails, is instead approximately exponential (Fig. 4a).

If we apply the method of sliding windows to the finance time series within the interval I shown in Fig. 2a, then we get Fig. 4b, which has artificially generated

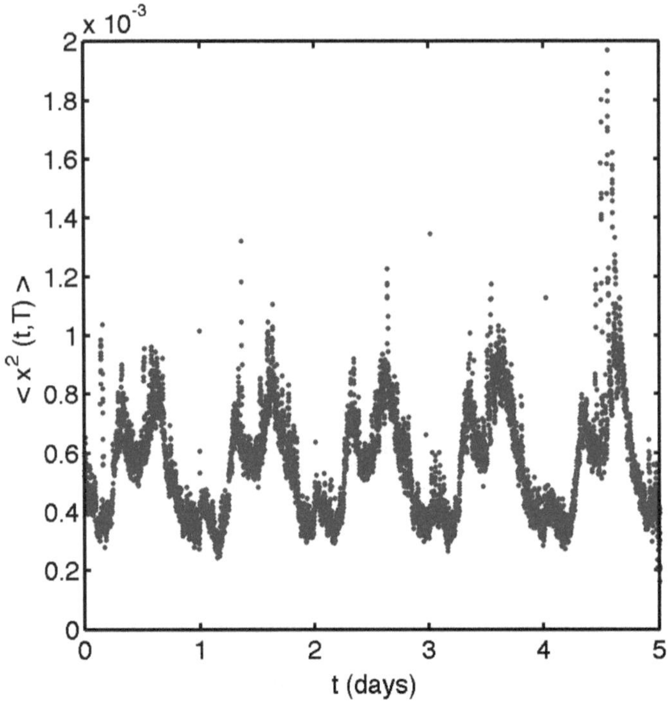

Fig. 2b We observe that the same intraday stochastic process occurs during each trading day. Both of the plots (a) and (b) would be flat were the increments x(t,T) stationary. Instead, the rms fluctuation of x(t,T) varies by a factor of 3 each day as t is varied, showing strong nonstationarity of increments. Note in (a) that we get scaling with H at best in the four disjoint colored regions, and with different values of H in each region (we find $H > 1/2$ where the colored slope is positive, otherwise $H < 1/2$)

fat tails and also a spurious Hurst exponent $H_S = 1/2$, just as with our numerical simulation using time series generated via the exponential model (Fig. 2a,b). *This shows how sliding windows can generate artificial fat tails and spurious Hurst exponents of 1/2 in data analysis.* That is, the use of sliding windows generates 'significant artifacts' when the increments are nonstationary. *Again, this is because sliding windows implicitly assume stationary increments.* This observation has far reaching consequences for the analysis of random time series, whether in physics, economics/finance, or biology.

A martingale process is indicated by intraday finance data for time intervals $T \geq 10$ min., and we end with a final note that the data analysis can be formulated systematically using the transition density characteristic of the underlying martingale. That transition density is the Green function for the drift-free Fokker-Planck pde with or without memory. Consider the data for one day as a run of the 'experiment'. Each day we start at a different initial return x_o, where typically $x_o \approx 0$. The increment mean square fluctuation for many independent runs (many different days)

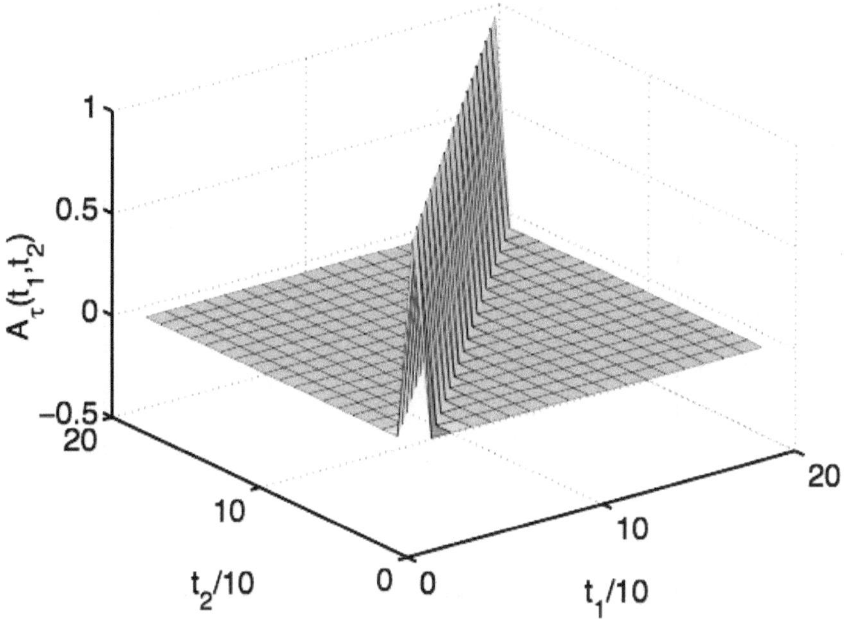

Fig. 3 Normalized autocorrelations in increments $A_T(t_1,t_2) = <x(t_1,T)x(t_2,T)> /(<x^2(t_1)> <x^2(t_2)>)^{1/2}$ for two nonoverlapping time intervals $[t_1,t_1+T]$, $[t_2,t_2+T]$ decay rapidly toward zero for $T \geq 10$ min. of trading

is then

$$\langle (x(t+T)-x(t))^2 \rangle = \int_{-\infty}^{\infty} (x-y)^2 g(x,t+T:y,t) f_1(x,t) dx dy \tag{28}$$

The diffusion coefficient is given by

$$D(x,t) \approx \frac{1}{T} \int_{-\infty}^{\infty} (x-y)^2 g(x,t+T;y,t) dx \tag{29}$$

for $0 < T \ll t$, so that

$$\langle (x(t+T)-x(t))^2 \rangle \approx T \int_{-\infty}^{\infty} D(y,t) f_1(y,t) dy \tag{30}$$

The scaling results used above agree with this prediction if $f_1(x,t)$ scales with H, requiring $f_1(x,t) = g(x,t;0,0)$, where the average over y is understood as an average over initial returns x_o for many separate runs/days. Ergodicity was not assumed, and cannot be assumed for nonstationary processes.

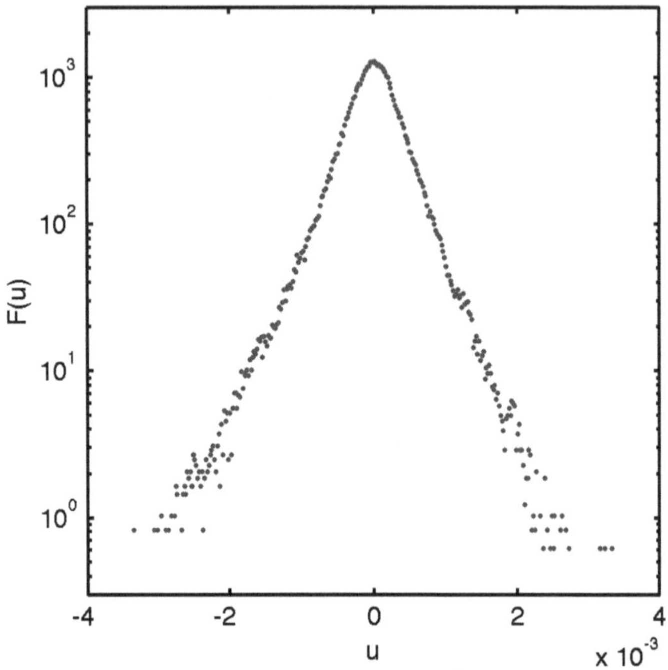

Fig. 4a Our scaling analysis focuses on the interval I shown in Fig. 2a. We plot the scaling function F(u) for H = 0.35 with 10 min. $\leq T \leq$ 160 min. Note that F(u) is slightly asymmetric and is approximately exponential, showing that the variance is finite

5 Shedding Fat from Cotton

Finally, consider Fig. 2 in Mandelbrot [19], where fat tails with infinite variance were deduced for cotton returns. He plots what he calls a 2nd moment, but which is analogous to the mean square fluctuation in our Fig. 2a and is simply our (30) above. Mandelbrot observes that quantity is 'badly behaved', doesn't 'converge', and assumes without proof that the cause is Levy-like fat tails (in a Levy density the variance is strictly infinite). He then set the 2nd moment equal to infinity, *assuming that the time series is stationary* so that his sliding window time average (our (30)) should make sense. But markets are nonstationary, are very far from statistical equilibrium, and in that case an ergodicity assumption about the empirical time average in (30) fails, the mean square fluctuation in (30) will not 'converge' but will fluctuate eradically if the increments are nonstationary. *The 'bad behavior' observed by Mandelbrot has nothing to do with fat tails and is instead direct evidence for nonstationarity of the increments.* His Fig. 2 is nothing more or less than the uneveness exhibited by noise traders like our Fig. 2a. We have produced evidence (Fig. 2b) that FX traders reproduce (at least before reading and being influenced by this paper) the same dynamics day after day, so the natural time scale for that analysis is one day. For cotton returns, the natural time scale for a correct data analysis

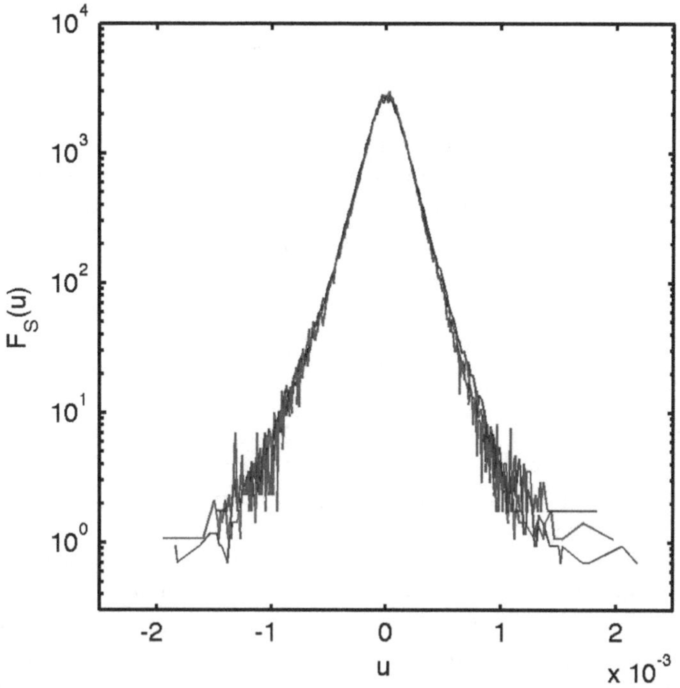

Fig. 4b The 'sliding interval scaling function' $F_s(u_s)$, $u_s = x_s(T)/T^{H_s}$, is constructed empirically from the same interval I for T = 10, 20, and 40 min. Note that fat tails have been generated spuriously by the sliding windows, and that a spurious Hurst exponent $H_s = 1/2$ has been generated as well, just as in the simulation data shown as Fig. 2a, b

is probably one year, with nonstationarity of increments reflecting unevenness of trading during the course of a year. Such seasonal variations cannot be smoothed out without masking the essence of the underlying market dynamics. It would be of interest to check cotton market returns for uncorrelated increments (to check for a martingale), where the diffusion coefficient (as explained above) would then describe the uneveness in the volatility of trading (the nonstationarity of the increments) over the time scale of a year. But a reliable cotton market analysis is made extremely difficult than FX because cotton price statistics are much more sparse, and will yield far more scatter in histograms than do FX market statistics where we cannot even get good enough daily returns histograms from 6 years of trading taken at 10 min. intervals. We would expect agricultural commodities in general to exhibit nonstationary increments.

6 The Efficient Market Hypothesis

Real finance markets are hard to beat, arbitrage possibilities are hard to find and, once found, are expected to disappear fast via arbitrage. The EMH is simply an attempt to mathematize the idea that the market is very hard to beat. If there is no useful information in market prices, then those prices can be counted as noise, the product of 'noise trading'. A martingale formulation of the EMH embodies the idea that the market is hard to beat, is overwhelmingly noise, but leaves open the question of hard to find correlations that might be exploited for exceptional profit. Our recent data analysis discussed above establishes (for Euro/Dollar data) that FX data over the past six years can be understood as a martingale in the log return variable x over the time interval of a day or a week, after the subtraction of a nearly trivial drift.

A strict interpretation of the EMH is that there are no correlations, no patterns of any kind, that can be employed *systematically* to beat the average return <R> reflecting the market itself: if one wants a higher return, then one must take on more risk (in the French-Fama way of thinking, 'omniscent agents' are assumed who neutralize all information up until time t_1). A Markov market is unbeatable, it has no systematically repeated patterns, no memory to exploit. We will argue below that the stipulation should be added that in discussing the EMH we should consider only normal, liquid markets, meaning very liquid markets with small enough transactions that approximately reversible trading is possible on a time scale of seconds [2, 40]. Otherwise, 'Brownian' market models do not apply. Liquidity, 'the money bath' created by the noise traders whose behavior is reflected in the diffusion coefficient, is somewhat qualitatively analogous to the idea of the heat bath in thermodynamics [2]: the second by second fluctuations in x(t) are created by the continual noise trading. When the liquidity dries up and the market crashes, as in the collapse of LTCM in 1998 or the subprime mortgage crisis of Fall, 2007, then Brownian models do not apply. In 1998 a lender of the last resort stepped in. Whether the lenders can provide the liquidity to prevent a depression in the subprime mortgage case is not known at the time of this article because hidden loses have yet to be reported. In any case the Dollar has sunk drastically since this lecture was given in April, and China, with the largest Dollar reserves in the world, now threatens to sell Dollars. That is, the Dollar has been systematically degraded as a currency and the end is not in sight.

Mandelbrot [22] proposed a less strict and very attractive definition of the EMH, one that directly reflects the fact that financial markets are hard to beat but leaves open the question whether the market can be beaten in principle at some high level of insight. He suggested that a martingale condition on returns realistically reflects the notion of the EMH. A martingale may contain memory, but that memory can't be easily exploited to beat the market precisely because the expectation of a martingale process x(t) at any later time is simply the last observed return. In addition, as we've shown above, pair correlations in increments cannot be exploited to beat the market either. The idea that memory may arise (in commodities, e.g.) from other variables (like the weather) [22] corresponds in statistical physics [42] to the appearance of memory as a consequence of averaging over other, more rapidly changing, variables in the larger dynamical system.

The martingale (as opposed to Markov) version of the EMH is also interesting because technical traders assume that certain price sequences give signals either to sell or buy. In principle, that is permitted in a martingale. A particular price sequence $(p(t_n),\ldots,p(t_1))$, were it quasi-systematically to repeat, can be encoded as returns $(x_n,\ldots,x_1)$ so that a conditional probability density $p_n(x_n;x_{n-1},\ldots,x_1)$ could be interpreted as a providing a risk measure to buy or sell. By 'quasi-repetition' of the sequence we mean that $p_n(x_n;x_{n-1},\ldots,x_1)$ is significantly greater than a Markovian prediction. Typically, technical traders make the mistake of trying to interpret random price sequences quasi-deterministically, which differs from our interpretation of 'technical trading' based on conditional probabilities (see Lo et al. [43] for a discussion of technical trading claims, but based on a non-martingale, non-empirically based model of prices). With only a conditional probability for 'signaling' a specific price sequence, an agent with a large debt to equity ratio can easily suffer the Gamblers' Ruin. In any case, we can offer no advice about technical trading, because the existence of market memory has not been firmly established (the question is left open by the analysis of [43]), liquid finance markets look pretty Markovian so far as we've been able to understand the data, but one can go systematically beyond the level of pair correlations to try to find memory. Apparently, this remains to be done, or at least to be published.

Memory could reflect heavy trading around a particular price and can, of course, be lost in the course of time. The writer remembers well the period of a few months ca. 1999 when CPQ sold for around \$22, and was traded often in the range \$18–25 before crashing further. Whether that provides an example is purely speculation at this point.

Fama [44] took Mandelbrot's proposal seriously and tried to test finance data at the simplest level for a fair game condition. We continue our discussion by first correcting a mathematical mistake made by Fama (see the first two of three unnumbered equations at the bottom of p. 391 in [44]), who wrongly concluded in his discussion of martingales as a fair game condition that $< x(t+T)\,x(t) >= 0$. Here's his argument, rewritten partly in our notation. Let x(t) denote a 'fair game'. With the initial condition chosen as $x(t_o) = 0$, then we have the unconditioned expectation $< x(t) >= \int x dx f_1(x,t) = 0$ (there is no drift). Then the so-called 'serial covariance' is given by

$$\langle x(t+T)x(t)\rangle = \int x dx < x(t+T) >_{cond(x)} f_1(x,t) \tag{31}$$

Fama states that this autocorrelation vanishes because $< x(t+T) >_{cond} = 0$. This is impossible: by a fair game we mean a Martingale, the conditional expectation is $< x(t+T) >_{cond} = \int y dy p_2(y,t+T;x,t) = x = x(t) \neq 0$, and so Fama should have concluded instead that $< x(t+T)\,x(t) >=< x^2(t) >$ as we showed in the last section. Vanishing of (31) would be true of statistically independent variables but is violated by a 'fair game'. Can Fama's argument be salvaged? Suppose that instead of x(t) we would try to use the *increment* $x(t,T) = x(t+T) - x(t)$ as variable. Then $< x(t,T)x(t) >= 0$ for a Martingale, as we showed in part 4. However, Fama's argument still would not be generally correct because x(t,T) canno*t* be taken as a 'fair game' variable unless the variance is linear in t, and in financial markets the variance is not linear in t. Fama's mislabeling of time dependent averages (typical in economics and finance literature) as ‘market equilibrium’ has been corrected elsewhere [2].

In our discussion of the EMH we shall not follow the economists’ tradition and discuss three separate forms (weak, semi-strong, and strong [45]) of the EMH, where a hard to test or effectively nonfalsifiable distinction is made between three separate classes of traders. We specifically consider only normal liquid markets with trading times at multiples of 10 min. intervals so that a Martingale condition holds [4]. Normal market statistics overwhelmingly (with high probability, if not ‘with measure one’) reflect the noise traders [3], so we consider *only* normal liquid markets and ask whether noise traders produce signals that one might be able to trade on systematically. The question whether insiders, or exceptional traders like Buffett and Soros can beat the market probably cannot be tested scientifically: even if we had statistics on such exceptional traders, those statistics would likely be too sparse to draw a conclusion. Furthermore, it is not clear that they beat liquid markets, some degree of illiquidity seems to play a significant role there. Effectively, or with high probability, there is only one type trader under consideration here, the noise trader. Noise traders provide the liquidity [2], their trading determines the form of the diffusion coefficient $D(x,t;\{x\})$, where $\{x\}$ reflects any finite memory present [8]. The question that we emphasize is whether, given a Martingale created by the noise traders, a normal liquid market can still be beaten systematically.

One can test for martingales and for violations of the EMH at increasing levels of correlation. At the level $n = 1$, the level of simple averages, the ability to detrend data implies a Martingale. At the level $n = 2$, vanishing increment autocorrelations implies a martingale. Both conditions are consistent with Markov processes and with the EMH. A positive test for a martingale *with memory* at the level $n \geq 3$ would eliminate Markov processes, and perhaps would violate the EMH as well. So far a we’re aware, this case has not yet been proposed or discussed in the literature. If such correlations exist and would be traded on, then a finance theorist would argue that they would be arbitraged away, changing the market statistics in the process. If true, then this would make the market even more effectively Markovian.

A Markov market cannot be systematically beaten, it has no memory of any kind to exploit. Volatility clustering and so-called ‘long term dependence’ appear in Markov models [32], are therefore not necessarily memory effects. In the folklore of finance it’s believed that some traders are able to make money from volatility clustering, which is a Markovian effect with a nontrivial variable diffusion coefficient $D(x,t)$, e.g. $D(x,t) = t^{2H-1}(1 + x/t^H)$.

(where x and t are absolute values) so one would like to see the formulation of a trading strategy based on volatility clustering to check the basis for that claim.

Testing the market for a non-Markovian martingale is nontrivial and apparently has not been done: tests at the level of pair correlations leave open the question of higher order correlations that may be exploited in trading. Whether the hypothesis of a martingale as EMH will stand the test of higher orders correlations exhibiting memory remains to be seen. In the long run, one may be required to identify a very liquid ‘efficient market’ as Markovian.

Acknowledgements First and foremost, I'm extremely grateful to my close colleagues Kevin Bassler and Gemunu Gunaratne. Excepting the historical part, this paper is based Second, I'm grateful to Harry Thomas for correcting errors in some of those papers, and to Roger Pynn for commenting critically on the historic part of this paper. Third and indispensable, my partner, sax playing wife, and local editor Cornelia Küffner read the lecture slides critically and improved the presentation. Those improvements are largely built into this paper.

References

1. B. Eichengreen, Globalizing Capital: A History of the International Monetary System, Princeton University Press, Princeton, NJ, 1998.
2. J.L. McCauley, Dynamics of Markets: Econophysics and Finance, Cambridge University Press, Cambridge, 2004.
3. K.E. Bassler, J.L. McCauley, & G.H. Gunaratne, Nonstationary Increments, Scaling Distributions, and Variable Diffusion Processes in Financial Markets, PNAS 104, 172287, 2007.
4. J.L. McCauley, G.H. Gunaratne, & K.E. Bassler, in Dynamics of Complex Interconnected Systems, Networks and Bioprocesses, eds. A.T. Skjeltorp & A. Belyushkin, Springer, New York, 2005.
5. R.L. Stratonovich, Topics in the Theory of Random Noise, tr. R.A. Silverman, Gordon & Breach, New York, 1963.
6. A.M. Yaglom & I.M. Yaglom, An Introduction to the Theory of Stationary Random Functions, tr. and eds. R.A. Silverman, Prentice-Hall, Englewood Cliffs, NJ, 1962.
7. R. Durrett, Brownian Motion and Martingales in Analysis, Wadsworth, Belmont, 1984.
8. J.L. McCauley, K.E. Bassler, & G.H. Gunaratne, Martingales, Detrending Data, and the Efficient Market Hypothesis, Physica A387, 202, 2008.
9. L. Arnold, Stochastic Differential Equations, Krieger, Malabar, 1992.
10. Christoph Hauert, in this book.
11. M.F.M. Osborne, The Random Character of Stock Market Prices, ed. P Cootner, MIT Press, Cambridge, MA, 1964.
12. C. Moore, Nonlinearity 4, 199, 727, 1991.
13. C.P. Kindleberger, Manias, Panics, and Crashes, A History of Financial Crises, Wiley, New York, 1996.
14. F. Black & M. Scholes, J. Pol. Econ. 8, 637, 1973.
15. P. Scholl-Latour, Das Schlachtfeld der Zukunft. Zwischen Kaukasus und Pamir, Goldman, München, 1998.
16. N. Dunbar, Inventing Money, Long-Term Capital Management and the Search for Risk-Free Profits, Wiley, New York, 2000.
17. P. Mirowski, Machine Dreams, Cambridge University Press, Cambridge, 2002.
18. F. Modigilani & M. Miller, The American Econ. Rev. XLVIII, 3, 261, 1958.
19. B. Mandelbrot, The Random Character of Stock Market Prices, ed. P. Cootner, MIT Press, Cambridge, MA, 1964.
20. K.E. Bassler, G.H. Gunaratne, & J.L. McCauley, Physica A 369, 343, 2006.
21. B. Mandelbrot & J.W. van Ness, SIAM Rev. 10, 2, 422, 1968.
22. B. Mandelbrot, J. Business 39, 242, 1966.
23. J.L. McCauley, G.H. Gunaratne, & K.E. Bassler, Hurst Exponents, Markov Processes, and Fractional Brownian Motion, Physica A 2007.
24. J.L. McCauley, G.H. Gunaratne, & K.E. Bassler, Martingale option pricing, Physica A 380, 351–356, 2007.
25. M.F.M. Osborne, The Stock Market and Finance from a Physicist's Viewpoint, Crossgar, Minneapolis, MN, 1977.
26. F. Black, J. Finance 3, 529, 1986.

27. E. Derman, My Life as a Quant, Wiley, New York, 2004.
28. P.L. Bernstein, Capital Ideas: The Improbable Origins of Modern Wall Street, The Free Press, New York, 1992.
29. M. Lewis, Liar's Poker, Penguin, New York, 1989.
30. S. Gallucio, G. Caldarelli, M. Marsili, & Y.-C. Zhang, Physica A 245, 423, 1997.
31. K.E. Bassler, to be submitted, 2007.
32. P. Embrechts & M. Maejima, Self-Similar Processes, Princeton University Press, Princeton, NJ, 2002.
33. A.L. Alejandro-Quinones, K.E. Bassler, M. Field, J.L. McCauley, M. Nicol, I. Timofeyev, A. Török, & G.H. Gunaratne, Physica A 363A, 383, 2006.
34. J.L. McCauley, Markov vs. nonMarkovian processes: A comment on the paper 'Stochastic feedback, nonlinear families of Markov processes, and nonlinear Fokker-Planck equations' by T.D. Frank, Physica A 382, 445–452, 2007.
35. P. Hänggi, H. Thomas, H. Grabert, & P. Talkner, J. Stat. Phys. 18, 155, 1978.
36. L. Borland, Phy. Rev. Lett. 89, 9, 2002.
37. C. Renner, J. Peinke, & R. Friedric, Physica A 298, 2001.
38. M.M. Dacoroga, G. Ramazan, U.A. Müller, R.B. Olsen, & O.V. Picte, An Introduction to High Frequency Finance, Academic, New York, 2001.
39. T. Di Matteo, T. Aste, & M.M. Dacorogna, Physica A 324, 183, 2003.
40. J.L. McCauley, K.E. Bassler, & G.H. Gunaratne, On the Analysis of Time Series with Nonstationary Increments in Handbook of Complexity Research, ed. B. Rosser, 2007.
41. K.E. Bassler, G.H. Gunaratne, & J.L. McCauley, Int. Rev. Fin. An., 2008 (to appear)
42. P. Hänggi & H. Thomas, Zeitschr. Für Physik B26, 85, 1977.
43. A. W. Lo, H. Mamaysky, & J. Wang, J. Finance LV, Nr. 4, 1705, 2000.
44. E. Fama, J. Finance 25, 383–417, 1970.
45. Johannes A. Skjeltorp, Scaling in the Norwegian stock market, Physica A 283, 486–528, 2000.

Evolutionary Dynamics of Genes and Environment in Cancer Development

Evolution of Cancer

Jarle Breivik

Abstract Cancer development involves dynamic interplay between genes and environment, and is increasingly understood as an evolutionary process within the body of an organism. Concurrently, human predisposition for cancer may also be regarded in an evolutionary perspective, shaped by interaction of genes and environment through the generations. We have developed a theoretical model that combines this somatic and germline evolution of cancer. Key predictions have been tested and confirmed by independent studies and the model has general implications for the understanding of cancer.

Keywords Cancer, evolution, genetics

1 Evolution Within

According to the current paradigm, cancer is a genetic disease caused by accumulation of mutations leading to breakdown of growth regulatory mechanisms [1, 2]. Concurrently, an equally important paradigm states that cancer is an environmental disease caused by factors like smoking, diet and radiation [3]. The connection between these contrasting perspectives is commonly perceived as a linear causation: Environments cause mutations and mutations cause cancer. Despite its apparent logic however, there is growing awareness that this level of reasoning represents an inadequate basis for modeling carcinogenesis. Genes and mutations evolve in dynamic interaction with the surrounding environment, and cancer development is increasingly understood as an evolutionary process within the organism [3, 4].

Jarle Breivik
Institute of Basic Medical Science, Faculty of Medicine
P.O. BOX 1018 Blindern, University of Oslo
0315 Oslo, Norway
E-MAIL: JBREIVIK@MEDISIN.UIO.NO

2 Cost and Benefit of DNA Repair

Different types of genetic instability are related to different biochemical environments within the body [5, 6]. Paradoxically, cancer cell lose the DNA repair mechanism that is protective in the given environment. To explain these findings we have proposed that mutagenic environments favor the loss of the repair and control mechanisms they activate [5, 7]. This model has been explicitly tested and confirmed by leading researchers in the field [8, 9].

The model implies competition between opposing genetic strategies in different environments and the basic logic is comprised by the phrase "Don't stop for repairs in a war zone" [10] (Fig. 1).

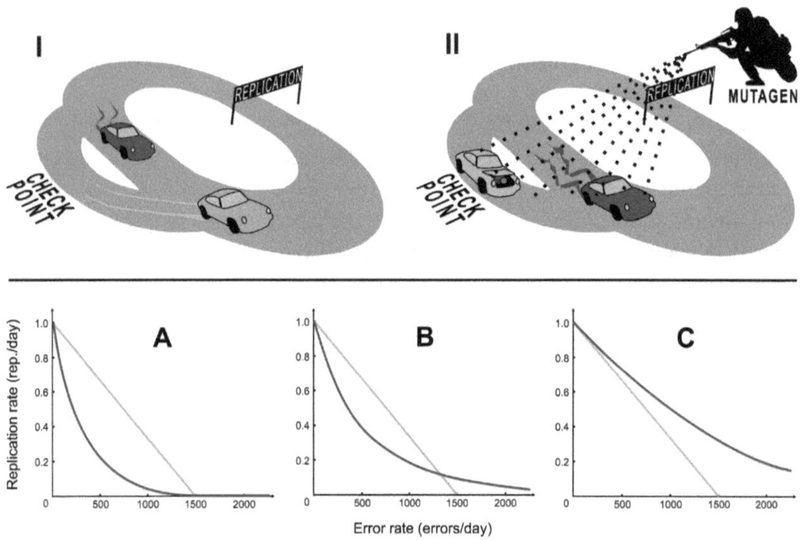

Fig. 1 Costs and benefits of repair. The problem of genetic instability may be viewed as an analogy to selecting repair strategies on a race track. **I**: A repair strategy (green) may be a winning tactic in amicable conditions. **II**: If errors appear faster then can they be removed, or for other reasons are impossible to repair, the repair strategy traps the vehicle in the check point. A repair deficient strategy (red) may then represent the better alternative. **A–C**: A repair strategy incurs a certain cost related to each repair process, whereas a repair deficient strategy incurs a cost related to the probability for unfavorable errors. The respective costs of the two strategies are therefore asymmetrical functions of mutation rate, and may be expressed in terms of replication rate. Replication rate with repair falls linearly with mutation rate until the repair capacity is exceeded, and replication comes to a complete stop (green lines). Replication rate without repair falls sharply at low, but flattens out at high mutation rates due to the probabilistic cost of errors (red lines). Combined there are three principle configurations depending on type of error, and efficiency of repair. **A**: The repair strategy is favorable at all viable mutation rates. **B**: The repair strategy is favorable at low, but not at high error rates. **C**: The repair strategy is unfavorable at all error rates. These configurations relate to the observations that some DNA repair mechanisms protect cells from mutagenic toxicity, whereas others involve increased susceptibility (Adapted from [10, 11])

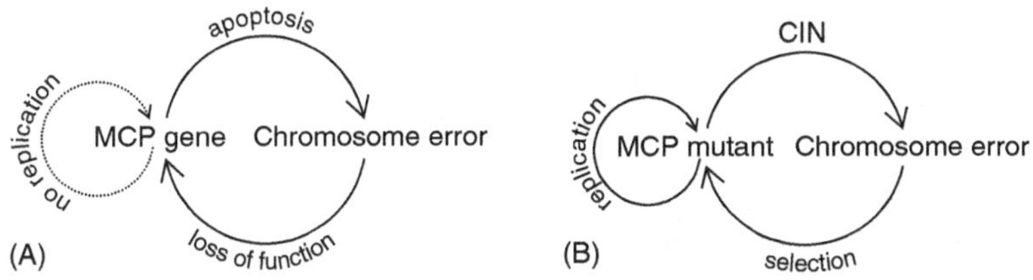

Fig. 2 Mitotic check point (MCP) genes block progression of the cell cycle in response to chromosome errors. Accordingly, such genes cannot replicate in the presence of a chromosome error. An environment that introduces chromosome errors will thus promote loss of mitotic check point genes and rise of chromosomal instability (CIN). The CIN phenotype then promotes more chromosomal errors, establishing a self-reinforcing cycle of instability (Adapted from [14])

3 Nonlinear Dynamics

Whereas the standard model of carcinogenesis involves a linear timeline of mutagenic events [1, 2], the theory of evolution implies a nonlinear relationship between genes and environment [12, 13]. The environment affects the genes, and the genetic system responds back on the environment. This principle may be illustrated by the well-known dynamics of antibiotic resistance: Antibiotics not only kill pathological bacteria, they also favor resistant variants by means of natural selection. A change in the environment favors an adaptive compensation in the genes, and a simple model of cause (antibiotic) and consequence (killing of bacteria) is inadequate.

Applied to cancer, we have used the reciprocal principles of evolution to explain the rise of chromosomal instability as an adaptive response to environmental mutagens (Fig. 2).

4 Germline vs. Somatic Evolution

Cancer development represents an evolutionary process within the body of an organism [15, 16], but is also related to the evolution of the genome through the generations [3]. The two processes are connected by the embryological division of germline and soma, and involve an intricate evolutionary dynamic of cooperation and competition [17] (Fig. 3).

5 Implications

Much research has focused on identifying the initiating factor of carcinogenesis. There are several conflicting theories and apparently no simple answer [4]. The evolutionary perspective redefines this problem and provides a very different solution.

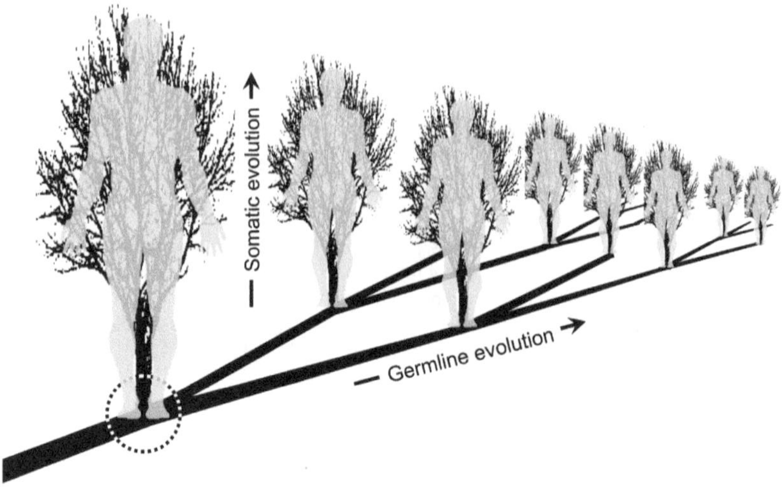

Fig. 3 Multicellular organisms imply a division between the somatic cells that make up the body and the germ cells that are propagated to the next generation. The genes of the germ line are selected for their ability to build successfully reproducing organisms, whereas genes of the somatic linage are selected for their ability to build successfully reproducing cells within the body of the organism. Accordingly, there is a conflict by which germline evolution favors collaboration, growth control and a high degree of genetic stability, whereas somatic evolution favors breakdown of cell cycle control and loss of genetic stability. This model provides a framework for integrating the somatic and germline evolution process, and has general implications for how to define the origin of cancer (Adapted from [14, 17])

The somatic lineage departs from the germline a few days after conception and is subject to mutation and selection throughout the life of the organism. A mutation may be present from birth but remains insignificant until it encounters a selective environment. Pinpointing an initiating factor may, thus, be neither possible nor relevant. Carcinogenesis is evolution by means of natural selection in the somatic environment, and its origin resides in our multicellular constitution.

References

1. Fearon ER, Vogelstein B. A genetic model for colorectal tumorigenesis. Cell 1990;61(5):759–67.
2. Arends JW. Molecular interactions in the Vogelstein model of colorectal carcinoma. J Pathol 2000 Mar;190(4):412–16.
3. Vineis P, Berwick M. The population dynamics of cancer: A Darwinian perspective. Int J Epidemiol 2006 Oct;35(5):1151–9.
4. Gibbs WW. Untangling the roots of cancer. Sci Am 2003 Jul;289(1):56–65.
5. Breivik J, Gaudernack G. Carcinogenesis and natural selection: A new perspective to the genetics and epigenetics of colorectal cancer. Adv Cancer Res 1999;76:187–212.
6. Breivik J, Gaudernack G. Carcinogenesis and natural selection: A new perspective to the genetics and epigenetics of colorectal cancer. Adv Cancer Res 1999;76:187–212.

7. Breivik J, Gaudernack G. Genomic instability, DNA methylation, and natural selection in colorectal carcinogenesis. Semin Cancer Biol 1999 Aug;9(4):245–54.
8. Bardelli A, Cahill DP, Kinzler KW, Vogelstein B, Lengauer C. Carcinogen-specific induction of genetic instability. Proc Natl Acad Sci U S A 2001;98(10):5770–5.
9. Herzog CR, Bodon N, Pittman B, Maronpot RR, Massey TE, Anderson MW, You M, Devereux TR. Carcinogen-specific targeting of chromosome 12 for loss of heterozygosity in mouse lung adenocarcinomas: Implications for chromosome instability and tumor progression. Oncogene 2004 Apr 15;23(17):3033–9.
10. Breivik J. Don't stop for repairs in a war zone: Darwinian evolution unites genes and environment in cancer development. Proc Natl Acad Sci U S A 2001 May 8;98(10):5379–81.
11. Breivik J, Gaudernack G. Resolving the evolutionary paradox of genetic instability: A cost-benefit analysis of DNA repair in changing environments. FEBS Lett 2004 Apr 9;563(1–3): 7–12.
12. Darwin CR. On the origin of species by means of natural selection. London: Murray, 1859.
13. Ridley M. Evolution, 3rd ed. Oxford: Blackwell Publishing, 2003.
14. Breivik J. The evolutionary origin of genetic instability in cancer development. Semin Cancer Biol 2005 Feb;15(1):51–60.
15. Cahill DP, Kinzler KW, Vogelstein B, Lengauer C. Genetic instability and Darwinian selection in tumours. Trends Cell Biol 1999 Dec;9(12):M57–M60.
16. Nowak MA, Michor F, Komarova NL, Iwasa Y. Evolutionary dynamics of tumor suppressor gene inactivation. Proc Natl Acad Sci U S A 2004;101(29):10635–8.
17. Breivik J. Cancer-evolution within. Int J Epidemiol 2006 Oct;35(5):1161–2.

Aging as Evolution-Facilitating Program and a Biochemical Approach to Switch It Off

Vladimir P. Skulachev

Abbreviations BLM, planar bilayer phospholipid membrane; C_{12}TPP, dodecyl triphenylphosphonium; DMQ, demethoxyMitoQL, MitoQ, compound of ubiquinone and decyl triphenylphosphonium; ROS, reactive oxygen species; SkQ, compounds of plastoquinone or methylplastoquinone and decyl (or amyl) triphenylphosphonium, methylcarninite, or tributylammonium; SkQ1, compound of plastoquinone and decyl triphenylphosphonium (other SkQ derivatives are shown in Fig. 3); $\Delta\psi$ transmembrane electric potential.

Abstract A concept is presented considering aging of living organisms as a final step of their ontogenetic program. It is assumed that such an aging program was invented by biological evolution to facilitate the evolutionary process. Indications are summarized suggesting that controlled production of toxic forms of oxygen (so called reactive oxygen species) by respiring intracellular organelles (mitochondria) is an obligatory component of the aging program. First results of a research project devoted to an attempt to interrupt aging program by antioxidants specifically addressed to mitochondria have been described. Within the framework of the project, antioxidants of a new type (SkQ) were synthesized. SkQs are composed of (i) plastoquinone (an antioxidant moiety), (ii) a penetrating cation, and (iii) a decane or pentane linker. Using planar bilayer phospholipid membranes, we selected SkQ derivatives of the highest penetrability, namely plastoquinonyl decyl triphenylphosphonium (SkQ1), plastoquinonyl decyl rhodamine 19 (SkQR1), and methylplastoquinonyl decyl triphenylphosphonium (SkQ3). Anti- and prooxidant properties of these substances and also of ubiquinonyl-decyl-triphenylphosphonium (MitoQ) were tested in isolated mitochondria. Micromolar concentrations of cationic quinones are found to be very strong prooxidants, but in

Vladimir P. Skulachev
Faculty of Bioengineering and Bioinformatics, Lomonosov Moscow State University, 119991 Moscow, Russia; fax: (495) 939–0338; E-mail: skulach@belozersky.msu.ru

Belozersky Institute of Physico-Chemical Biology, Lomonosov Moscow State University, 119991 Moscow, Russia the lower (sub-micromolar) concentrations they display antioxidant activity which decreases in the series SkQ1 = SkQR1 > SkQ3 > MitoQ. Thus, the window between the anti- and prooxidant effects is the smallest for MitoQ and the largest for SkQ1 and SkQR1. SkQ1 is rapidly reduced by complex III of the mitochondrial respiratory chain, i.e. it is a rechargeable antioxidant. Extremely low concentrations of SkQ1 and SkQR1 completely arrest the H_2O_2-induced apoptosis in human fibroblasts and HeLa cells (for SkQ1, $C_{1/2} = 8 \cdot 10^{-9}$ M). Higher concentrations of SkQ1 are required to block necrosis initiated by reactive oxygen species (ROS). In mice, SkQ1 decelerates the development of three types of accelerated aging (progeria) and also of normal aging, and this effect is especially demonstrative at early stages of aging. The same pattern is shown in invertebrates (*Drosophila* and *Daphnia*), and fungus (*Podospora anserina*). In mammals, the effect of SkQs on aging is accompanied by inhibition of development of such age-related diseases as osteoporosis, involution of thymus, cataract, retinopathy, etc. SkQ1 manifests a strong therapeutic action on some already pronounced retinopathies, in particular, congenital retinal dysplasia. With drops containing 250 nM SkQ1, vision is recovered in 66 of 96 animals (dogs, cats and horses) who became blind because of retinopathy. SkQ1-containing drops instilled into eyes prevent the loss of sight in rabbits suffering from experimental uveitis and restore vision to animals that had already become blind due to this pathology. A favorable effect is also achieved in experimental glaucoma in rabbits. Moreover, the pretreatment of rats with 0.2 nM SkQ1 significantly decreases the H_2O_2-induced arrhythmia of the isolated heart. SkQ1 strongly reduces the damaged area in myocardial infarction or stroke and prevents the death of animals from kidney infarction. In $p53^{-/-}$ mice, SkQ1 decreases the ROS level in the spleen cells and inhibits appearance of lymphomas which are the main cause of death of such animals. As a result, the lifespan increases. SkQs look like promising drugs to treat aging and age-related diseases.

Keywords Biological evolution, aging, mitochondria, reactive oxygen species, SkQs, antioxidants

1 Aging as a Program

In multicellular organism, death of a cell is as a rule a result of operation of suicide programs encoded in the cell genome. These programs are different types of apoptosis and necrosis or their combinations (for reviews, see [1, 2]). Suicide programs are also found in bacteria [3] and unicellular eukaryotes [4]. This means that unicellular organisms possess programs for self-elimination. Phenomena of such a kind (we have dubbed "phenoptosis" [1]) are also inherent in multicellular organisms [2, 5, 6], but their molecular mechanisms are still to be elucidated.

The proof of the genetically programmed death of individuals has markedly reinforced positions of a few gerontologists who, following the great biologists of the second half of the nineteenth century—Darwin [7], Wallace, and Weissmann [8]—are inclined to consider aging of an organism as a final stage of ontogeny and not a simple consequence of accumulation of accidental errors.

If aging is programmed, a reasonable question arises what the biological role of this program is. We suggest that this role is to accelerate evolution. Natural selection can simply not notice the appearance of a small new potentially useful change unessential for survival of a young and strong animal. However, this change may become decisive for an organism weakened by aging. Take such an example. Two young hares, a clever one and a silly one, on meeting a fox have nearly similar chances to escape from the enemy because they can run much faster. But with age the clever hare will get an advantage compared to the silly one because the speed of the hares' running will decrease as a result of an age-related disease known as sarcopenia (diminution of cell number in muscles). And now the clever hare who will immediately take to his heels on seeing the fox will have much better chance to save his life than the silly hare who lingers at the start. Thus, only the clever hare will continue procreating leverets, and, as a result, the hare population will become wiser [2].

Modern humans stopped adapting themselves to the environment. Instead of adapting, they try to change the environment to their benefit. If we want to fly, we build aircraft rather than wait until wings will grow on our back. Therefore, mechanisms specialized for acceleration of evolution are atavistic for human and should be abolished if these mechanisms, such as aging, are obviously harmful for an individual. And in accordance with this logic, we may speak about the control of the aging program as a most important problem of medicine of the future.

2 What Does the Aging Program Consist of?

The free radical hypothesis by Harman [9] is the most popular concept on the causes of aging. Harman supposed that oxidation of biopolymers by reactive oxygen species plays a leading role in weakening of vital functions with age. This hypothesis has been supported by the finding that the oxidation level of DNA, proteins, and lipids increases with age. Such a situation can be a consequence of an increase in the reactive oxygen species (ROS) production or a decrease in the antioxidant defense in old age, or simply a result of the prolonged damaging action of ROS cumulated proportionally to the organism's age [1, 2, 5, 6]. In the framework of the theory of programmed aging, Harman's hypothesis has to be supplemented by an assumption that an increase in the ROS-induced damages with time is controlled by the organism, similarly to any other change occurring during ontogenesis.

But what is the role of ROS generation in triggering aging processes? The cell contains a set of enzymes that convert O_2 to the primary form of ROS: superoxide ($O_2^{-\cdot}$) or its derivative hydrogen peroxide (H_2O_2) [2, 10]. However, the power of all these enzymes is significantly lower than the power of enzymes of the inner mitochondrial membrane respiratory chain. Mitochondria of the human adult take up about 400 liters of O_2 per day and convert it to water (H_2O) by four-electron reduction. However, if even 0.1% of this amount of O_2 is reduced through the chemically easier one-electron pathway, this will produce 0.4 liter of $O_2^{-\cdot}$, which exceeds the abilities of all other mechanisms of ROS generation taken together.

In the respiratory chain, $O_2^{-\cdot}$ is generated mainly in complexes I and III. The highest rate of superoxide generation in intact mitochondria is observed in complex I on the reverse transfer of electrons from succinate to NAD^+ at the maximum proton potential (i.e. in the absence of ADP). This rate equal to about 1 nmol O_2/mg protein per minute is nearly fivefold higher than the rate of $O_2^{-\cdot}$ production in complex III under the same conditions. It is 10% of the respiration rate without ADP (in state 4) and about 1.5% of the maximal respiration rate in the presence of ADP (in state 3). Even if this rate is lowered because of the NAD^+ present [11], a rather impressive amount of superoxide is generated considering the high toxicity of $O_2^{-\cdot}$ conversion products. In fact, we carry in our mitochondria a potential generator of strong toxin, which can easily kill our cells and us together with them. And this catastrophe will be caused not only by the direct toxic action of ROS, but as a consequence of triggering apoptosis and necrosis, which are induced by ROS.

In 2007, Lambert et al. [12] reported that the lifetime of mammals and birds is longer the lower is the rate of H_2O_2 generation by heart mitochondria during the reverse electron transfer from succinate through complex I (animals of 12 species very differently systematically positioned were studied: from mice to cattle and baboons and from quails to pigeons). No correlation was found when H_2O_2 generation was measured which was coupled to forward electron transfer (through both complex I and complex III). The longest-lived African rodent, the naked mole-rat, seems to be the only exception to this rule. This creature of a mouse size is famous for its lifespan of about 28 years instead of the mouse's 2.5–4 years. Mole-rat mortality does not depend on age [13] as if the aging program is absent in this animal. The program seemed to be switched off downstream of ROS because both the rate of ROS generation at the reverse electron transfer [12] and the level of biopolymer oxidation [14, 15] in mole-rats are higher than in mice. However, the latter finding is not surprising on taking into account that activities of superoxide dismutase and catalase in these two species are not very different, and the third antioxidant enzyme, glutathione peroxidase, is 70-fold less active in mole-rats than in mice [16]. The key for understanding this paradoxical situation seems to be presented by the observation that even very high concentrations of H_2O_2 failed to induce apoptosis in the mole-rat's cells [17]. In this respect, the mole-rat is reminiscent of long-lived mouse strain having mutation in genes encoding the protein p66shc [18], the enzyme of CoQ synthesis mClk1 [19], or elongation factor 2 kinase [20] whose cells are also resistant to apoptogenic action of hydrogen peroxide [18, 20] or the prooxidant menadione [19].

All these observations are in good agreement with our hypothesis on the structure of the aging program (Fig. 1) suggesting a chain of events from "biological clock" which controls ontogenesis, through generation of mitochondrial ROS, to diminution in the cell number in organs due to apoptosis triggered by these ROS, and, as a result, to weakening of the organism's vital functions [2].

Ontogenetic clock

↓

Mitochondrial ROS

↓

Apoptosis

↓

Diminution of cell number in organs

↓

Weakening of organ functions

Fig. 1 Hypothetical mechanism of aging mediated by mitochondrial ROS [2]

3 Antioxidants as Geroprotectors

Based on this scheme, it can be asserted that antioxidant protection of cells against mitochondrial ROS has to arrest or at least decelerate aging. As to the treatment of aging with antioxidants, the literature is rather vast and ambiguous: from the statement of Ames and colleagues that such a drug against age is already found [21, 22] to conclusions of Howes about a complete fruitlessness of this approach and, thus, erroneousness of Harman's hypothesis [23].

We think that there are two essential omissions in the works concerning the treatment of aging with antioxidants.

(1) If it is necessary to neutralize just intramitochondrial ROS generated by complex I, antioxidants addressed to mitochondria are required. However, no attempts to use such antioxidants in gerontology are described in the literature. The statement of B. N. Ames [22] that the action of *tret*-butylhydroxylamine used in his experiments is addressed to mitochondria is absolutely speculative. A rather high effective concentration (10^{-4} M) of this substance seems to indicate that the Ames' suggestion is erroneous. In our experiments, effective concentrations of a mitochondria-addressed geroprotector SkQ1 are much lower ($5 \cdot 10^{-10}$–10^{-5} M, see below).

(2) Commonly used antioxidants, even if they, like vitamin E, can reach the mitochondrial membrane (just as any other cellular membranes) are *natural* substances, and their excess can be cleaved by cellular enzymes if their presence is undesirable for carrying out the ontogenetic program. We are dealing with just such a situation when vitamin E is used to arrest aging. In spite of increased vitamin E uptake, the organism tries to realize the aging program as a final step of ontogenesis. To this end, the coming vitamin E is cleaved by cytochrome P450, which is induced in the liver by excess of this vitamin. In fact, the organism disposes systems for defense against both oxygen and antioxidants. The matter is that ROS are

responsible for some biological functions indispensable for a normal life. They are directly involved in the struggle with bacteria and viruses and also indirectly help to overcome in this struggle because they are required, in particular, for functioning of the immune system, where ROS-triggered apoptosis plays an important role [2]. A vital role of ROS are proved by direct experiment carried out by Goldstein. Mice and rats were kept in air deprived of superoxide, and in three weeks the animals died [24].

Thus, a "superantioxidant" intended to block the aging program and thus to prolong a healthy life has to meet the following requirements: (1) the antioxidant must not remove ROS completely but only their excess generated inside mitochondria at the reverse electron transfer in the respiratory chain; (2) the antioxidant must not be inactivated by enzymes of the organism which is by all means seeking to complete its ontogenesis by carrying out the aging program.

4 Mitochondria-Targeted Antioxidants

Both above problems might be solved based on the phenomenon observed in our group in cooperation with the group of E. A. Liberman already in 1969–1970 [25, 26, 27, 28]. We mean the discovery of *penetrating ions*, i.e. hydrophobic synthetic compounds that can easily penetrate across the mitochondrial membrane in spite of presence of an ionized group in their structure. Triphenylmethylphosphonium proved to be a typical representative of such ions. In this cation, the positive charge on the phosphorus atom is strongly delocalized over three phenyl residues. As a result, water dipoles cannot be held on the cation and do not form an aqueous shell preventing penetration of the ion into the hydrophobic membrane core. The mitochondrial interior is the only intracellular compartment negatively charged relative to its environment (i.e. the cytosol); therefore, on entering the cell, penetrating cations will specifically be accumulated within mitochondria. This accumulation can be described by the Nernst equation (a 10-fold gradient at $\Delta\psi = 60$ mV). If in the energized mitochondria $\Delta\psi$ is $\sim$180 mV [29], the penetrating cation concentration in the mitochondrial matrix has to be 1,000 times higher than in the cytosol.

This reasoning allowed us to suppose that penetrating cations can be used by mitochondria as "electric locomotive molecules" for accumulation of uncharged substances attached to these cations [30]. Such an idea allowed to explain the role of cationic group of carnitine in the transport of fatty acid residues into mitochondria [29, 30]. On being electrophoretically concentrated in the inner leaf of the phospholipid bilayer of the inner mitochondrial membrane, fatty acid acyls and carnitine would be able to act as antioxidants interrupting chain reactions of peroxidation of phospholipids and proteins which constitute this leaf. This hypothesis has been confirmed by data obtained in our group by Yu. N. Antonenko and A. A. Pashkovskaya. It was found that palmitoyl carnitine prevented photooxidation of gramicidin incorporated into a plane bilayer phospholipid membrane (BLM, unpublished observation).

At border of the two centuries, the principle of "electric locomotive molecule" was used by M. P. Murphy for the addressed delivery into mitochondria of antioxidants, namely vitamin E [31] and ubiquinone [32]. The so-called MitoQ composed of ubiquinone and decyl triphenylphosphonium cation seemed to be especially promising. MitoQ is obviously more advantageous than palmitoyl carnitine or cationic derivative of vitamin E because its reduced form oxidized during its functioning as antioxidant can be regenerated by accepting of electrons from the respiratory chain. In other words, MitoQ is a *rechargeable* antioxidant [33]. In fact, in the experiments of Murphy et al. MitoQ was accumulated and reduced by mitochondria and protected them and also cell cultures against oxidative stress [31–37].

We have confirmed the data of Murphy et al. on the antioxidant activity of MitoQ. However, we found that it became prooxidant with an increase in its concentration. Thus, for rat heart mitochondria $C_{1/2}$ of the MitoQ antioxidant effect and prooxidant effect was 0.3 and 0.5 μM, respectively (experiments by M. Yu. Vyssokikh and coauthors, our group). Addition of 1.5–2.5 μM MitoQ caused such a fast generation of hydrogen peroxide by mitochondria oxidizing NAD-dependent substrates that its rate was higher than all recorded values described in the literature and approximated the respiration rate in state 4 [38]. Thus, MitoQ is hardly the best mitochondria-targeted antioxidant. Therefore, we decided to start a search for antioxidants stronger than MitoQ and with a broader window between the anti- and prooxidant effects.

5 Starting the Project on Search of "Superantioxidant"

In 2003, we started with the project on using penetrating ions in practice. It was initiated owing to grants from the Charitable Foundation "Paritet" (now "Volnoe Delo") established by O. V. Deripaska. O.V.D. was graduated from the Physical Faculty of Moscow State University (MSU) in 1993. Now he is one of the richest businessmen in Russia. O.V.D. is a member of the MSU Board of Trustees and significantly helps to his *Alma Mater*. In early 2003, I shared with him a plan of creating a "superoxidant" drug with the bold hope of stopping aging. I did not promise the personal immortality to the foundation's creator, but mentioned that even if we fail to beat aging, we will at least help to overcome hangnails. O.V.D. burned with the idea that aging was a harmful atavism which might be cancelled. He granted us $120,000 per year for five years.

Upon obtaining the money, we stood before the usual dilemma of what to buy: a fish or a fishing-rod. We preferred the rod and ordered synthesis of demethoxyMitoQ (DMQ) because the nematode *Caenorhabditis elegans* was shown to live sevenfold longer if it has deletions in the gene encoding an enzyme which converts demethoxyCoQ to ubiquinone and in the insulin receptor gene [39]. The substance

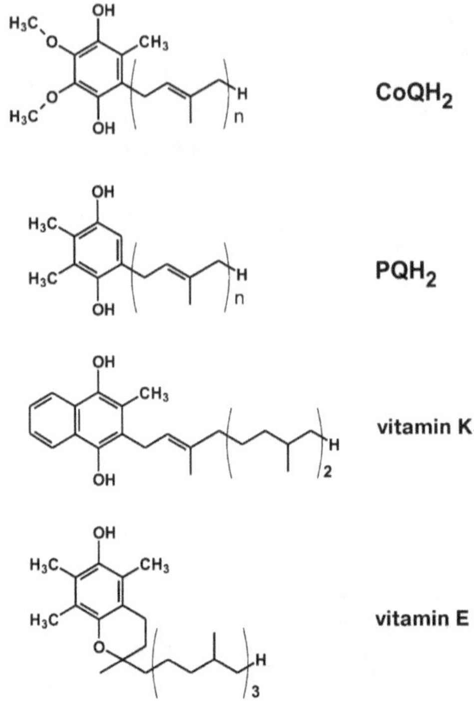

Fig. 2 Some quinone derivatives operating as electron carriers and/or antioxidants

was synthesized and displayed an increased difference between the anti- and prooxidant concentrations compared to MitoQ, but unfortunately this difference was less pronounced than we would have liked.

Then we paid attention to *plastoquinone*, an electron carrier acting instead of ubiquinone in photosynthetic electron transport chains in the plant chloroplasts as well as in cyanobacteria. During evolution, ubiquinone involved in the mitochondrial respiratory chain was substituted by plastoquinone in the chloroplast chain of the same plant cell, possibly, just because of the better antioxidant properties of plastoquinone, as shown in chemical experiments on model systems [40, 41]. In fact, a chloroplast generating oxygen is under conditions of much stronger oxidative stress than a mitochondrion which takes up oxygen. Plastoquinone, in contrast to ubiquinone, has methyl groups instead of methoxy groups. As to the ubiquinone methyl group, it is substituted in plastoquinone by hydrogen (Fig. 2).

Whereas the anti- and prooxidant concentrations of MitoQ differ less than twofold (300 and 500 nM), this difference for a plastoquinone derivative of decyl triphenylphosphonium named SkQ1 was found to increase to 32-fold (25 and 800 nM) [38].

This result showed that SkQ1 is a very effective mitochondria-addressed antioxidant possessing at low concentrations no prooxidant effect. In this connection, I proposed Mr. Deripaska to transform our grant to an investment project aimed at creation of a new type of drugs and biotechnology products on the base of SkQ. The proposal was accepted, and the investment project was started in 2005.

The equipment (sometimes very expensive) was bought, laboratories were repaired, and new researchers were invited; some of them returned from abroad to Moscow. The project rapidly left the limits of the Belozersky Institute of Physico-Chemical Biology where the work was performed in the framework of the initial grant. At present, the project includes groups from several MSU faculties and research institutes in Moscow, St. Petersburg; Novosibirsk, Rostov as well as in Johnson Medical Institute (USA) and in Venner-Gren Institute (Sweden).

6 SkQs: *In Vitro* Experiments

On searching for the best antioxidant among cationic derivatives of quinones, a number of substances were synthesized mainly by our chemists G. A. Korshunova, N. V. Sumbatyan, and L. S. Yaguzhinsky. It was taken into account that (1), in contrast to ubiquinone (CoQ), plastoquinone and another chloroplast electron carrier vitamin K_1, as well as a "professional antioxidant" vitamin E, contain no methoxy groups, and (2) vitamin K_1, vitamin E, and CoQ have in the fifth position of the quinone ring a methyl group, which is absent in plastoquinone (Fig. 2). As a result, the majority of our quinone derivatives contained either plastoquinone (SkQ1, SkQ2M, SkQ4, SkQ5, SkQR1) or methylplastoquinone (SkQ3). For comparison, MitoQ and DMQ containing, respectively, ubiquinone and demethoxyubiquinone, were also synthesized.

As cations, first of all alkyl triphenylphosphonium derivatives were used with the alkyl residue linking the cation with the quinone being either decyl (SkQ1, SkQ3, SkQ4, DMQ, and MitoQ) or amyl (SkQ5). In some cases a different cation was used. Instead of phosphonium, compounds with an ionized nitrogen atom were tried: methylcarnitine (SkQ2M), decyltributylammonium (SkQ4), or Rhodamine 19 (SkQR1, Fig. 3).

All the prepared compounds were first tested for their ability to penetrate across model membranes. I. I. Severina showed that SkQR1 and SkQ3 display the best penetrability across a BLM. The gradient of their concentrations generated Nernst diffusion potential with the "plus" sign in the compartment with the lower concentration of the cation in the range $5 \cdot 10^{-6}$–$5 \cdot 10^{-5}$ M (SkQR1) or $5 \cdot 10^{-6}$–$5 \cdot 10^{-4}$ M (SkQ3) (SkQR1 at concentrations higher than $5 \cdot 10^{-5}$ M damaged the BLM). The corresponding concentrations of SkQ1, MitoQ, and DMQ were in a higher range ($5 \cdot 10^{-5}$–$5 \cdot 10^{-4}$ M) indicating a slightly lower permeability of the BLM for these compounds. The BLM permeability for SkQ2M, SkQ4, and SkQ5 was still lower, so that the Nernst potential values could not be reached even in the presence of tetraphenylborate increasing the BLM cationic permeability [38]. Based on these findings, SkQ1, SkQR1, and SkQ3 were chosen for further work.

Anti- and prooxidant properties of these substances were studied in our group by M. Yu. Vyssokikh et al. who compared them with those of MitoQ, DMQ, and

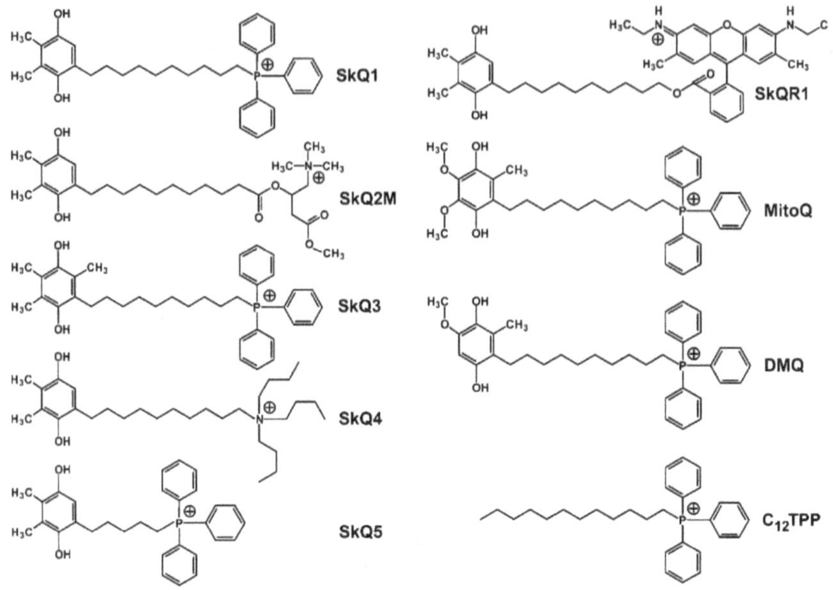

Fig. 3 Some substances that were synthesized and used in the project (the reduced forms are presented) [38]

C_{12}TPP (dodecyl triphenylphosphonium, i.e. an SkQ analog lacking quinone). To measure the antioxidant activity of these compounds, rat heart mitochondria were energized by oxidizing succinate and incubated under conditions of the Fenton reaction when $OH^{\cdot}$ radicals were generated from hydrogen peroxide produced by the mitochondria (ascorbic acid was used as the reductant of added iron ions). Under these conditions, mitochondrial phospholipids (first of all, cardiolipin) were peroxidized, and this peroxidation was followed by accumulation of malonic dialdehyde. All substances tested, except C_{12}TPP, inhibited this accumulation, but their effective concentrations were very different (Table 1). SkQ1 and SkQR1 were the most active. The antioxidant activity of others decreased in the following series: DMQ > SkQ3 > MitoQ. The activity of decylplastoquinone was nearly

Table 1 Effect of SkQ1, SkQR1, and MitoQ on H_2O_2-induced cell death from apoptosis and fragmentation of mitochondria in human fibroblasts

	Cell death		Fragmentation of mitochondria	
Quinone	$C_{1/2}$, M	Maximum protection, %	$C_{1/2}$, M	Maximum protection, %
SkQ1	1×10^{-9}	100	1×10^{-10}	70
SkQR1	1×10^{-12}	100	2×10^{-12}	100
MitoQ	1×10^{-8}	85	2×10^{-8}	62

100-fold, and that of MitoQ was more than 30-fold, lower than the activity of SkQ1.

The prooxidant activity was tested (i) by the rate of non-enzymatic one-electron reduction of O_2 to $O_2^{-\cdot}$ mediated by hydroquinone forms of the corresponding compounds and (ii) by the stimulation of H_2O_2 production by rat heart mitochondria oxidizing glutamate and malate in state 4. An important result of the first test was the several times slower reduction of O_2 by $SkQ1H_2$ than by $MitoQH_2$ (data by M. Yu. Vyssokikh and E. K. Ruuge). In the second test, these compounds displayed no obvious difference. However, it should be noted that the prooxidant effect of MitoQ on the mitochondria increased at slightly lower concentrations than that of SkQ1 and SkQ3 [38].

The further experiments on cells and animals were performed with SkQ1 and SkQR1, which manifested the best antioxidant properties and a good penetrability when tested, respectively, on mitochondria and BLM. Due to a high quantum yield of the SkQR1 fluorescence, its fate could be traced in the cells and the organism. In HeLa cells, SkQR1 specifically stained only the mitochondria and on intraperitoneal injection into animals it was mainly accumulated in the kidneys and liver (data of B. V. Chernyak's and D. B. Zorov's groups).

Data of Yu. M. Vassiliev, B. V. Chernyak et al., showed that extremely low concentrations of SkQ1 and SkQR1 prevented the death of human cells in culture in the presence of low concentrations of H_2O_2. Fragmentation of long mitochondria into small ones (the "thread–grain transition" [42]) in fibroblasts and HeLa cells, induced by H_2O_2 and other apoptogens, was shown to be the most sensitive parameter. This transition could be prevented by 2-h preincubation with SkQR1, SkQ1, and MitoQ at the $C_{1/2}$ value, respectively, of $2 \cdot 10^{-12}$, $1 \cdot 10^{-10}$, and $2 \cdot 10^{-8}$ M. SkQ1 very effectively prevented apoptosis of these cells ($C_{1/2} = 1 \cdot 10^{-9}$ M, for MitoQ $C_{1/2} = 1 \cdot 10^{-8}$ M; the antioxidants were added seven days before the addition of H_2O_2). As shown in Table 1, MitoQ was not only one order of magnitude less active, but, even at the optimal concentration, it failed to completely prevent the fragmentation of mitochondria and cell death, whereas SkQR1 and SkQ1 did. This failure of MitoQ could not be overcome by increasing its level. At concentrations higher than $1 \cdot 10^{-7}$ M, MitoQ, similarly to SkQ derivatives, it did not weaken but actually strengthened the apoptogenic effect of hydrogen peroxide.

The efficiency of such low concentrations of SkQ1 and SkQR1 in experiments on cells is explained by some factors. First, the coefficient of their distribution between the hydrophobic and aqueous phases is very high: in the octanol/water system, it is 13,000 for SkQ1. Then, we have gained three orders more due to effect of $\Delta\psi$ (~180 mV) on the inner mitochondrial membrane and one order more from the $\Delta\psi$ on the outer cell membrane (~60 mV). In total, the SkQ1 concentration gradient between the extracellular medium and inner half-membrane layer of the inner mitochondrial membrane occurs to be really immense: $13{,}000 \cdot 1000 \cdot 10 = 1.3 \cdot 10^8$. This means that at the SkQ1 concentration in the medium of $1 \cdot 10^{-10}$ M (Table 1), its concentration in the inner half-membrane layer will be $1 \cdot 10^{-10} \cdot 1.3 \cdot 10^8 = 1.3 \cdot 10^{-2}$ M. On the way to the inner mitochondrial membrane, SkQ1 crosses the outer cell membrane, endoplasmic reticulum membranes, and the mitochondrial outer membrane accumulating within them in accordance with the lipid/water distribution coefficient. That is why a prolonged preincubation is required to provide for the protective effect of picomolar concentrations of SkQ1 on the intracellular mitochondria. The equilibrium concentration in the system of medium/intracellular mitochondria will require less time the higher is the penetrability of our cations. Therefore, in experiments with a relatively short (for 2 h) preincubation with SkQR1 was markedly more efficient than with SkQ1.

The very high efficiency of SkQs is also explained by their being rechargeable antioxidants. Thus, SkQ1 can be easily reduced by complex III of the respiratory chain [38].

And one more reason has to be taken into account, namely, the mechanism itself of the effect of SkQs on H_2O_2-induced apoptosis. In our experiments, $(1–4) \cdot 10^{-4}$ M H_2O_2 used for triggering apoptosis induced a dramatic pulse in the generation of endogenous ROS by mitochondria, and this seemed to be a signal for the cell to commit suicide. Upon saturating the inner half-membrane layer of a mitochondrion, SkQ1 interrupts in the very beginning the chain peroxidation of fatty acid residues of phospholipids. It seems that this function needs no large amount of the antioxidant. More likely, the antioxidant is highly available for hydrophobic regions of the membrane where these residues are located. Such availability depends on the orientation of SkQs within the membrane: the charged phosphonium group has to be in water, and the flexible decane linker terminated with plastoquinone has to be in the membrane depth. The SkQ molecule can move along the membrane and the plastoquinone head can oscillate pendulum-like between the membrane core and surface.

In line with the above-described logic, the effective antioxidant concentrations of SkQ1 on the isolated mitochondria under conditions of external generation of $OH^{\cdot}$ radicals occurred to be higher than on the cells ($C_{1/2} = 2.5 \cdot 10^{-8}$ M, data by M. Yu. Vyssokikh [38]) instead of $1 \cdot 10^{-9}$ M $- 1 \cdot 10^{-10}$ M on the cells (see Table 1). Still greater amounts of SkQ1 were required to arrest necrosis of cells induced by generation of radicals on a photosensitizer (MitoTracker Red) under the influence of light ($C_{1/2} = 3.5 \cdot 10^{-7}$ M) after preincubation for 1 h with SkQ1. In this case, MitoQ was also less effective than SkQ1 (data of D. S. Izyumov and B. V. Chernyak [38]). It should be noted that highly efficient prevention by cationic quinones of endogenous oxidative stress in cell cultures was for the first time observed by Murphy et al. on fibroblasts from patients with Friedreich's ataxia [37]. This disease is associated with disorders in production of the mitochondrial protein frataxin, which results in an increase in the content of iron ions in the matrix and in oxidative stress. Such fibroblasts can survive in culture only in the presence of antioxidants, in particular, decylubiquinone ($C_{1/2} = 3 \cdot 10^{-8}$ M). MitoQ was much more effective than decylubiquinone ($C_{1/2} = 5 \cdot 10^{-10}$ M), and an uncoupler increased this value to $1 \cdot 10^{-8}$ M. An excess of MitoQ ($5 \cdot 10^{-7}$ M and more) could not protect the cells against oxidative stress, whereas an excess of decylubiquinone ($1 \cdot 10^{-4}$ M) was as efficient as its lower quantities.

7 SkQs as Antioxidant Promising for Treatment of Age-Related (and Not Only Age-Related) Diseases

On the next stage of the work, the antioxidant effect was tested on isolated rat heart. V. I. Kapelko and V. L. Lakomkin (Research Center of Cardiology) induced severe heart arrhythmia by addition of H_2O_2 into the perfusate. The arrhythmia was significantly less pronounced if the hearts had been isolated from rats pretreated daily for two weeks with SkQ1 ($2 \cdot 10^{-10}$ mol/kg body weight) [43].

Then effect of SkQ1 on the *in vivo* development of heart pathology was studied. O. I. Pisarenko et al. (Cardiology Research Center) used SkQ1 for a similar pretreatment of rats before the artificial induction of myocardial infarction. SkQ1 ($2.5 \cdot 10^{-7}$ mol/kg) decreased by 40% the damaged zone of the heart muscle [43]. In our institute, D. B. Zorov and E. Yu. Plotnikov removed one kidney of rats and some time later induced infarction of the remained kidney by interrupting the blood flow for 90 min. This resulted in the death of 80% of the experimental rats five-six days after the ischemization. However a single intraperitoneal injection of 0.5–1 μmol SkQ1 or SkQR1 per kilogram body weight the day before ischemization decreased the mortality to 20% [43]. A similar pretreatment with SkQR1 sharply (threefold) decreased the zone of brain necrosis in rats with artificial stroke (data by D. B. Zorov and N. K. Isaev). Injected in this manner, SkQ1 was inefficient, possibly because of its lower penetrability [43].

The majority of *in vivo* studies in the framework of the project were designed to answer the main question of whether SkQs could serve as geroprotectors. Effects of SkQ1 were studied in three models of accelerated aging (progeria) and in normal aging of a rodent, drosophila, daphnia and a fungus. Positive effects of SkQ1 were observed in X-ray-induced progeria of mice (A. G. Ryazanov et al., Johnson Medical Institute, USA) and in progeria caused by a point mutation in the gene of mitochondrial DNA polymerase (I. Shabalina, B. Cannon, and J. Niedergard, Venner-Gren Institute, Sweden). In the second case, mice with mutated mitochondrial DNA polymerase unable to correct its own mistakes were used (in the mutant mice an aspartate in the proof-reading exonuxlease domain of the enzyme was substituted by alanine [44]). A 90% mutant mice not treated with SkQ1 died by the 280th day of the life with all signs of premature aging (body weight loss, osteoporosis causing lordokyphosis, decrease in body temperature and inability to maintain it in the cold, alopecia, torpor, etc.). Only 10% in the group of the mice treated with SkQ1 (5 μmol/kg) died during this period. Most signs of aging in these animals were poorly pronounced or even absent [43].

Many experiments were conducted on rats of the OXYS strain with progeria caused by permanent oxidative stress. The selection of this strain was initiated by R. I. Salganik at the Novosibirsk Institute of Cytology and Genetics [45]. The strain is characterized by development, already a few months after birth, of such age-related diseases as cataract, retinopathy, osteoporosis, thymus involution, decreased sex motivation in males, and memory worsening [46–48]. N. G. Kolosova and colleagues in the same institute found that the development of all these signs of aging was strongly decelerated by treatment with SkQ1 (50–250 nmol/kg per day with food). Eye drops containing $2.5 \cdot 10^{-7}$ M SkQ1 cured the already developed cataract and retinophy within 1.5 months. This effect was confirmed by both photographs of the eyeground (data by N. G. Kolosova and A. Zh. Fursova) and electron microscopy (data by L. E. Bakeeva and V. B. Saprunova). A similar treatment of cataract was also favorable in Wistar rats [43]. Wistar rats seldom live till retinophy appears, thus we had to change the object. Researchers of the Skryabin Veterinary Academy E. P. Kopenkin and L. F. Sotnikova jointly with I. I. Senin et al from our institute obtained a therapeutic effect of SkQ1 drops in retinopathies in dogs, cats, and horses. In total, 122 animals who failed to react to routine treatments were treated with SkQ1. The most clear positive effect was recorded in congenital retinal dysplasia (a radical improvement in 67% of cases) and in secondary degeneration of the retina (54%). In progressing retinal degeneration, SkQ1 was less effective (29%). In 96 cases, our patients were completely blind animals, and 66 of them began to see upon daily treatment with SkQ1 drops ($2.5 \cdot 10^{-7}$ M) during two-to-six weeks. Such a favorable SkQ1 effect was confirmed by measuring electroretinograms [43].

These positive results are in contrast with the failure of researchers of Murphy's group who tried to treat with MitoQ congenital retinopathies of mice [49]. This may be explained by the closeness of the antioxidant and prooxidant concentrations of MitoQ [38]. However, the authors assumed [49] that the studied retinopathies cannot be cured by antioxidants at all.

To conclude this series of studies, we have turned to artificially induced eye disease of animals, in particular, to experimental uveitis (I. I. Senin's group in our institute) and glaucoma (Helmholtz Institute, V. P. Erichev and coworkers). Both these diseases are known to be associated with strong oxidative stress [50, 51], and in both cases SkQ1 drops gave favorable results [43]. Experiments with autoimmune uveitis induced by immunization of rabbits with a photoreceptor cell protein arresting, and resulting in blindness were especially demonstrative. Into one eye of the blind animals, drops of $2.5 \cdot 10^{-7}$ M SkQ1 were instillated four times a day, and several days later the vision of the treated eye recovered whereas the non-treated eye retained blind. The same procedure prevented the development of uveitis if the drops were given shortly after the immunization. P. P. Philippov from our institute together with E. N. Grigoryan (Koltsov Institute of Developmental Biology) succeeded in modeling an aspect of the SkQ1 effect in the treatment of retinopathies: in roller cultures of eye fundus, SkQ1 decreased eightfold the transformation of the pigmented epithelium cells of the retina into phagocytes that could destroy this tissue [43]. In the Petrov Institute of Oncology (St. Petersburg), the group of the Russian Gerontological Society President V. N. Anisimov studied effect of SkQ1 on aging of normal mice without any form of progeria. SkQ1 (0.5, 5, or 50 nmol/kg) noticeably decreased the age-related mortality of the animals, and the effect was especially demonstrative on the first 20% of the deceased animals (see below, Fig. 4c). The lifetime of these animals was increased 2.5-fold. For 50% of the deceased animals the effect was 30%, and for the long-lived mice (the last 20% of the cohort studied) it was only about 15%. In other words, the so-called rectangularization of

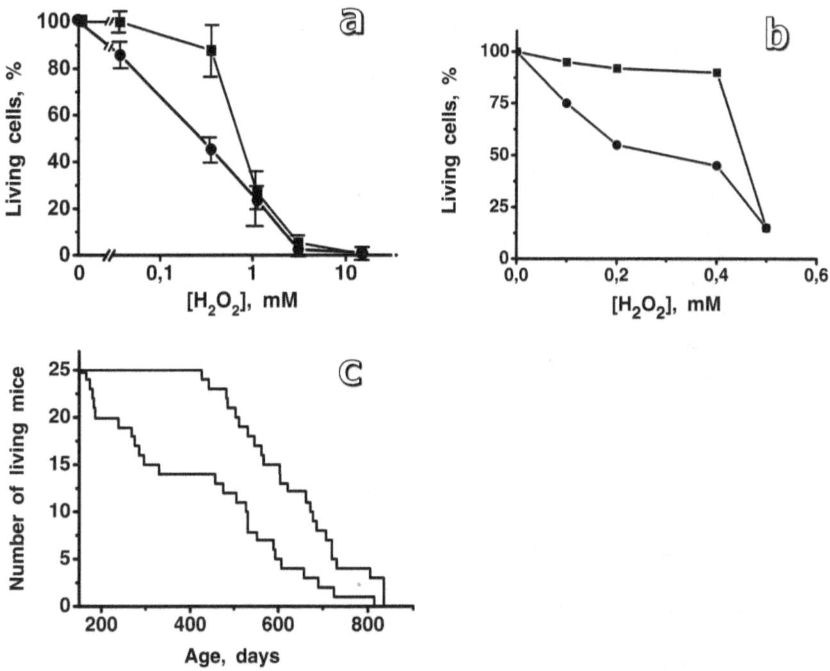

Fig. 4 Rectangularization of survival curves of yeast cells (a) [52], human fibroblasts (b) [38], and mice (c) [43] upon switching off the programs of cell suicide and aging. The lower curves present control, the upper curves show switching off of the programs by cycloheximide (a) or SkQ1 (b and c)

the mortality curve took place. The same change in the mortality curve was revealed for invertebrates – *Drosophila* (data of E. G. Pasyukova et al., Institute of Molecular Genetics), or *Daphnia* (data of O. F. Filenko, Faculty of Biology, MSU); as well as for a fungus (*Podospora anserina*, data of M. Yu. Vyssokikh). In all these cases, the median lifespan was increased in the presence of SkQ1.

The effect of SkQ1 on lifetime was associated with improvement of the life quality. In particular, regular estrous cycles lost in 70% of mice by the 22nd month of life were similar to those in young mice if mice received SkQ1. The antioxidant dramatically changed the pattern of causes of death. Thus, SkQ1 strongly decreased mortality from infections, which increased in untreated mice because of the aging-associated weakening of immunity. Mortality from some types of tumors was also decreased. However, SkQ1 was impotent against mammary gland cancer which became the main cause of death of the animals treated with SkQ1 (the experiment was conducted on females). SkQ1 had virtually no influence on mortality of the HER-2 mice predisposed to mammary gland cancer. These mice lived less than one year (as compared with the lifespan of normal mice which was much more than two years under the same conditions) [43].

Table 2 SkQs decelerate aging of animals

1	Decrease age-dependent mortality and increase the lifespan
2	Decelerate development of retinopathies and display therapeutic effect on
3	already developed retinopathies
4	Decelerate development of cataract and display therapeutic effect on
5	already developed cataract
6	Decelerate development of uveitis and display therapeutic effect on
7	already developed uveitis
8	Decelerate development of glaucoma
9	Decrease myocardial infarction zone
10	Abolish heart arrhythmia
11	Decrease stroke zone
12	Prevent death in kidney infarction
13	Decelerate age-related degeneration of thymus
14	Decelerate development of lymphomas and some other tumors in p53-
15	lacking animals
16	Decelerate loss of sexual motivation in males
17	Prevent the loss of estrous cycle in females
18	Stimulate healing of wounds
19	Decelerate osteoporosis
20	Decelerate alopecia in mice
21	Decelerate loss of whiskers in mice
22	Decelerate graying in radiation-induced progeria
23	Increase level of NO that can decelerate development of hypertension
24	Decelerate development of progeria in mitochondrial DNA polymerase mutants

In experiments of B. P. Kopnin, M. P. Lichinitser et al. (Russian Oncology Center, Moscow), very low doses of SkQ1 (5 nmol/kg) significantly increased lifetime of mice deprived of p53 protein and as a result suffering from oxidative stress. These mice die mainly because of lymphomas by the 250th day of life. The survival curve of such mice drinking an SkQ1 solution was noticeably shifted to the right, along with a pronounced decrease in the ROS level in their spleen cells [43]. It was shown in the same laboratory that the human intestinal carcinoma cells lacking p53 produced tumors much faster on grafting to athymic mice. This increased aggressiveness of the carcinoma cells was completely abolished in the SkQ1-drinking mice (50 nmol/kg) [43].

Our data on the influence of SkQ1 and SkQR1 on animals *in vivo* are summarized in Table 2. One can see that 24 different traits of aging are mitigated under the influence of SkQs, and this allows us to characterize our substances as powerful geroprotectors. Moreover, the geroprotective effect of SkQ1 is specified not only by an increase in the maximum lifetime but also by improvement of the quality of life during the second half of the lifespan. In other words, it prolongs youth.

It is interesting to compare the effect of SkQ1 on the mouse survival, its effect on fibroblast death caused by H_2O_2, and the effect of cycloheximide on the H_2O_2-induced death of yeast (Fig. 4). No doubt the latter two cases represent a blockade of the cell suicide program: in the fibroblasts it is blocked due to prevention by

SkQs of the effect of ROS on the mitochondrial membrane, and in the yeast this blockade is caused by switching off synthesis of proteins necessary for phenoptosis. Note that in all the three cases the mortality curve is similarly changed, namely it is rectangularized. At high H_2O_2 (about 0.5 mM), the cells seem to die independently of the apoptotic or phenoptotic mechanisms but simply because of a direct toxic action of this substance. Therefore, inhibition of the suicide programs failed to save the cells. It is likely that in the case of treatment of mice with SkQs we are dealing with a complete inhibition of the aging program, and the animals die because the organism is worn out and has accumulated too many breakages during the long life.

Another possibility consists in that at very old age an aging program sensitive to low doses of SkQ can be replaced by another program that is insensitive to such doses. Just such a variant seems to be realized in drosophila in which the geroprotective effect of SkQs is observed only during the very early stages of aging and then disappears. If the hypothesis about two aging programs is also true for mammals, the lifetime in the late ages can be prolonged either by increasing the dose of SkQs or by combination of SkQ with biguanide type geroprotectors, which act mainly in the end of life [53]. Corresponding experiments are now commenced in the group of V. N. Anisimov. Concurrently, the effect of SkQ1 on the lifetime of the eEF2 kinase knock-out mice is under study in the laboratory of A. G. Ryazanov. The lifetime of these mice is longer due to a decreased late mortality [20]. We are also going to test a p66shc$^{-/-}$ mutant [18] whose mortality curve is similar to that of the above-mentioned mutant.

The above findings indicate that SkQ1 itself is very promising for creating a new generation of drugs acting on aging-related processes as an antioxidant specifically addressed to mitochondria. On the other hand, the action of SkQs is not limited to their geroprotective effect. Experiments with heart arrhythmia, myocardial and kidney infarctions, stroke, wound healing, tumor inoculation, etc. were performed on young animals. It might be that SkQs simply "purify the dirtiest place in the cell" (the mitochondrial interior) that results in a nonspecific favorable effect on very different aspects of the vital activity of cells. However, this explanation is rather unlikely. In such a case, it remains unclear why, e.g., plants synthesizing separately plastoquinone and penetrating cation berberine did not create during evolution their combination capable of removing ROS from the interior of plant mitochondria. As was already noted, these mitochondria are subjected to even stronger "oxygen danger" than mitochondria of animals because the cytosol is saturated with the chloroplasts-generated oxygen.

There is an alternative: the program used for triggering aging as a slow phenoptosis can be also used in other fatal programs inducing a fast, or acute, phenoptosis. And here we stand in front of the fundamental question of biology: how genetic programs can exist which are counterproductive for an individual.

Even in the nineteenth century Ch. Darwin [7], A. P. Wallace, and A. Weissmann [8] supposed that death of an individual can be altruistic being useful for family or community. In 1964, W. D. Hamilton published a series of two articles entitled "Genetic Evolution of Social Behavior" [54]. In 1976, the book "The Selfish Gene" [55] by R. Dowkins appeared, where the author developed and popularized the idea of Hamilton that not a species, group, or even an individual is the main unit of selection, but that the gene is such a unit. In essence, here we are dealing not with the well-being of a community but with dictatorship of the genome, which is the only self-reproducing biological structure, and its retention, development, and expansion has priority compared to the well-being of an individual or a group of individuals. According to this concept, an organism is only a construction, a machine, providing for the interests of the genome.

Some years ago I formulated the so-called "Samurai Law of Biology—It is better to die than to be wrong", or in more extended form, "Complex biological systems (from organelles and higher) are equipped with programs of self-elimination which are actuated when this system becomes dangerous for any other system of higher position in the biological hierarchy" [2]. Combined with the concept about genome dictatorship, this law means that any critical state of the organism when it is unable to ensure the safety of its genome and in the case of convalescence can generate offspring with the altered genome has to be a signal for the organism's self-elimination, i.e. phenoptosis. As spoke Monsieur Bahys, a good-for-nothing physician in Molière's comedy "L'Amour médecin" ("Love as a Healer")—"Il vaut mieux mourir selon les règles, que de réchapper contre les règles" ("It is better to die according to rules than to recover against the rules").

It seems probable that mechanisms of rapid phenoptosis and slow phenoptosis (aging) are mediated by intramitochondrial ROS during early stages of the process. If this hypothesis is true, the favorable effect of SkQs not only in aging but in very different acute diseases of young organisms may be explained by quenching of these ROS.

It is likely that SkQs can be used as a tool in "rebellion of machines"—in the attempt of *Homo sapiens* to put an end to genome tyranny and to cancel those of the genome-dictated programs that are useful for the genome but disadvantageous for the individual. Acute phenoptosis of a human in a critical state and terminating his existence by biochemical suicide according to the above-mentioned "Bahys principle" seems to be as harmful an atavism as aging. Possibly, there are other genetic programs counterproductive for the organism. And their cancellation would symbolize a conversion of humans to *Homo sapiens discatenatus* (from Latin *catena*—chains, irons) that would be a highest achievement of the twenty-fist century medicine.

Acknowledgements I would like to express my sincere gratitude to leaders of the groups participating in the Project V. N. Anisimov, Yu. N. Antonenko, L. E. Bakeeva, B. Cannon, B. V. Chernyak, O. F. Filenko, E. N. Grigorian, V. I. Kapel'ko, N. G. Kolosova, E. P. Kopenkin, B. P. Kopnin, G. A. Korshunova, M. P. Lichinitser, I. V. Manukhov, E. N. Mokhova, M. S. Muntyan, E. G. Pasyukova, P. P. Philippov, E. I. Rogaev, E. K. Ruuge, A. G. Ryazanov, I. I. Senin, I. I. Severina, M. V. Skulachev, L. F. Sotnikova, V. N. Tashlitsky, Yu. M. Vassiliev, M. Yu. Vyssokikh, L. S. Yaguzhinsky, D. B. Zorov, to MSU Rector and Chairman of the Advisory Board V. A. Sadovnichii and also to the Board members O. S. Vikhansky M. P. Kirpichnikov and V. A. Tkachuk, to leaders of "RAInKo" and "Mitotechnology" companies S. V. Karabut and A. A. Grigorenko, to the "Ikontri" Consulting company Chief V. V. Perekhvatov and his coworkers I. V. Skulachev and K. V. Skulachev, to Deputy Director of Belozersky Institute V. A. Drachev, to my Secretary O. O. Malakhovskaya. I would like to emphasize that the Project could be realized due to the generous financial support by O. V. Deripaska, to whom I am very thankful.

This report is mainly based upon my paper published in Biochemistry (Moscow) 72, 1700–1714, 2007.

References

1. Skulachev, V. P. (1999) *Biochemistry (Moscow)*, **64**, 1418–1426.
2. Skulachev, V. P. (2003) in *Topics in Current Genetics. Model Systems in Ageing* (Nystrom, T., and Osiewacz, H. D., eds.) Vol. 3, Springer-Verlag, Berlin-Heidelberg, pp. 191–238.
3. Lewis, K. (2000) *Microbiol. Mol. Biol. Rev.*, **64**, 503–514.
4. Longo, V. D., Mitteldorf, J., and Skulachev, V. P. (2005) *Nature Rev. Genet.*, **6**, 866–872.
5. Skulachev, V. P. (2005) *Vestn. Ros. Akad. Nauk.*, **75**, 831–843.
6. Skulachev, V. P., and Longo, V. D. (2005) *Ann. N. Y. Acad. Sci.*, **1057**, 145–164.
7. Darwin, Ch. (1871) *The Descent of Man*, Murray, London.
8. Weissmann, A. (1889) *Essays upon Heredity and Kindred Biological Problems*, Clarendon Press, Oxford.
9. Harman, D. (1956) *J. Gerontol.*, **11**, 298–300.
10. Skulachev, V. P. (1999) *Mol. Asp. Med.*, **20**, 139–184.
11. Grivennikova, V. G., and Vinogradov, A. D. (2006) *Biochim. Biophys. Acta*, **1757**, 553–561.
12. Lambert, A. J., Boysen, H. M., Buckingham, J. A., Yang, T., Podlutsky, A., Austad, S. N., Kunz, T. H., Buffenstein, R., and Brand, M. D. (2007) *Aging Cell*, **6**, 607–618.
13. Buffenstein, R. (2005) *J. Gerontol. Biol. Sci.*, **60**, 1369–1377.
14. Andziak, B., and Buffenstein, R. (2006) *Aging Cell*, **5**, 525–532.
15. Andziak, B., O'Connor, T. P., Qi, W., DeWaal, E. M., Pierce, A., Chaudhuri, A. R., van Remmen, H., and Buffenstein, R. (2006) *Aging Cell*, **5**, 463–471.
16. Andziak, B., O'Connor, T. P., and Buffenstein, R. (2005) *Mech. Ageing Dev.*, **126**, 1206–1212.
17. Labinsky, N., Csiszar, A., Orosz, Z., Smith, K., Rivera, A., Buffenstein, R., and Ungvari, Z. (2006) *Am. J. Physiol. Heart Circ. Physiol.*, **291**, H2698–H2704.
18. Migliaccio, E., Giorgio, M., Mele, S., Pelicci, G., Revoldi, P., Pandolfi, P. P., Lanfrancone, L., and Pelicci, P. G. (1999) *Nature*, **402**, 309–313.
19. Liu, X., Jiang, N., Hughes, B., Bigras, E., Shoubridge, E., and Hekimi, S. (2006) *Gen. Dev.*, **19**, 2424–2434.
20. Chu, H.-P., Grigorian, I. A., Dorovkov, M. V., Nagele, R. G., Komarova, E. A., Gudkov, A. V., Harrison, D. E., and Ryazanov, A. G. (2008) *Nature*, in press.
21. Hagen, T. M., Liu, J., Lykkesfeldt, J., Wehr, C. M., Ingersoll, R. T., Vinarsky, V., Bartholomew, J. C., and Ames, B. N. (2002) *Proc. Natl. Acad. Sci. USA*, **99**, 1870–1875.
22. Atamna, H., Robinson, C., Ingersoll, R., Elliott, H., and Ames, B. N. (2001) *FASEB J.*, **15**, 196–204.
23. Howes, R. M. (2006) *Ann. N. Y. Acad. Sci.*, **1067**, 22–26.
24. Goldstein, N. (2002) *Biochemistry (Moscow)*, **67**, 161–170.
25. Liberman, E. A., Topali, V. P., Tsofina, L. M., Jasaitis, A. A., and Skulachev, V. P. (1969) *Nature*, **222**, 1076–1078.
26. Grinius, L. L., Jasaitis, A. A., Kadziauskas, Yu. L., Liberman, E. A., Skulachev, V. P., Topali, V. P., Tsofina, L. M., and Vladimirova, M. A. (1970) *Biochim. Biophys. Acta*, **216**, 1–12.
27. Bakeeva, L. E., Grinius, L. L., Jasaitis, A. A., Kuliene, V. V., Levitsky, D. O., Liberman, E. A., Severina, I. I., and Skulachev, V. P. (1970) *Biochim. Biophys. Acta*, **216**, 12–21.
28. Liberman, E. A., and Skulachev, V. P. (1970) *Biochim. Biophys. Acta*, **216**, 30–42.
29. Skulachev, V. P. (1988) *Memb. Bioenerg.*, Springer-Verlag, Berlin.
30. Severin, S. E., Skulachev, V. P., and Yaguzinsky, L. S. (1970) *Biokhimiya*, **35**, 1250–1257.
31. Smith, R. A., Porteous, C. M., Coulter, C. V., and Murphy, M. P. (1999) *Eur. J. Biochem.*, **263**, 709–716.

32. Kelso, G. F., Porteous, C. M., Coulter, C. V., Hughes, G., Porteous, W. K., Ledgerwood, E. C., Smith, R. A., and Murphy, M. P. (2001) *J. Biol. Chem.*, **276**, 4588–4596.
33. Murphy, M. P., and Smith, R. A. (2007) *Annu. Rev. Pharmacol. Toxicol.*, **47**, 629–656.
34. James, A. M., Cocheme, H. M., Smith, R. A., and Murphy, M. P. (2005) *J. Biol. Chem.*, **280**, 21295–21312.
35. Kelso, G. F., Porteous, C. M., Hughes, G., Ledgerwood, E. C., Gane, A. M., Smith, R. A., and Murphy, M. P. (2002) *Ann. N. Y. Acad. Sci.*, **959**, 263–274.
36. Saretzki, G., Murphy, M. P., and von Zglinicki, T. (2003) *Aging Cell*, **2**, 141–143.
37. Jauslin, M. L., Meier, T., Smith, R. A., and Murphy, M. P. (2003) *FASEB J.*, **17**, 1972–1974.
38. Antonenko, Yu. N., Archipova, L. T., Archipova, M. M., Bakeeva, L. E., Chernyak, B. F., Domnina, L. V., Fursova, A. Zh., Grigorian, E. N., Ivanova, O. Yu., Izyumov, D. S., Khailova, L. S., Klishin, S. S., Kolosova, N. G., Kopenkin, E. P., Korshunov, S. S., Korshunova, G. A., Kovaleva, N. A., Lyamzaev, K. G., Muntyan, M. S., Nepryakhina, O. K., Pashkovskaya, A. A., Philippov, P. P., Pletjushkina, O. Yu., Pustovidko, A. V., Rokitskaya, T. I., Ruuge, E. K., Saprunova, V. B., Senin, I. I., Severina, I. I., Simonyan, R. A., Skulachev, I. V., Skulachev, M. V., Sotnikova, L. F., Sumbatyan, N. V., Tashlitsky, V. N., Trofimova, N. A., Vassiliev, Yu. M., Vyssokikh, M. Yu., Yaguzhinsky, L. S., and Skulachev, V. P. (2008), in press.
39. Lakowski, B., and Hekimi, S. (1996) *Science*, **272**, 1010–1013.
40. Kruk, J., Jemiola-Rzeminska, M., and Strzalka, K. (1997) *Chem. Phys. Lipids*, **87**, 73–80.
41. Roginsky, V., Barsukova, T., Loshadkin, D., and Pliss, E. (2003) *Chem. Phys. Lipids*, **125**, 49–58.
42. Skulachev, V. P., Bakeeva, L. E., Chernyak, B. V., Domnina, L. V., Minin, A. A., Pletjushkina, O. Yu., Saprunova, V. B., Skulachev, I. V., Tsyplenkova, V. G., Vasiliev, J. M., Yaguzhinsky, L. S., and Zorov, D. B. (2004) *Mol. Cell. Biochem.*, **256/257**, 341–358.
43. Skulachev, V. P., et al., in preparation.
44. Trifunovic, A., Wreeenberg, A., Falkenberg, M., Spelbrink, J. N., Rovio, A. T., Bruder, C. E., Bohlooly, Y. M., Gidlof, S., Oldfors, A., Wilbom, R., Tornell, J., Jacobs, H. T., and Larsson, N.-G. (2004) *Nature*, **429**, 417–423.
45. Solov'eva, N. A., Morozkova, T. S., and Salganik, R. I. (1975) *Genetika*, **11**, 63–71.
46. Kolosova, N. G., Lebedev, P. A., Aidagulova, S. V., and Morozkova, T. S. (2003) *Bull. Exp. Biol. Med.*, **136**, 415–419.
47. Sergeeva, S., Bagryanskaya, E., Korbolina, E., and Kolosova, N. (2006) *Exp. Gerontol.*, **41**, 141–150.
48. Kolosova, N. G., Shcheglova, T. V., Sergeeva, S. V., and Loskutova, L. V. (2006) *Neurobiol. Aging*, **27**, 1289–1297.
49. Vlachantoni, D., Tulloch, B., Taylor, R. W., Turnbull, D. M., Murphy, M. P., and Wright, A. F. (2006) *Invest. Ophthalmol. Vis. Sci.*, **E-5773**.
50. Rajendram, R., Saraswathy, S., and Rao, N. A. (2007) *Br. J. Ophthalmol.*, **91**, 531–537.
51. Moreno, M. C., Campanelli, J., Sande, P., Sanez, D. A., Keller Sarmiento, M. I., and Rosenstein, R. E. (2004) *Free Rad. Biol. Med.*, **37**, 803–812.
52. Madeo, F., Frohlich, E., Ligr, M., Grey, M., Sigrist, S. J., Wolf, D. H., and Frohlich, K.-U. (1999) *J. Cell Biol.*, **145**, 757–767.
53. Anisimov, V. N., Semenchenko, A. V., and Yashin, A. I. (2003) *Biogerontology*, **4**, 297–307.
54. Hamilton, W. D. (1964) *J. Theor. Biol.*, **7**, 1–16, 17–52.
55. Dowkins, R. (1976) *The Selfish Gene*, Oxford University Publishers, Oxford.

Evolution of Vision

Mikhail Ostrovsky

Abstract The evolution of photoreception, giving rise to eye, offers a kaleidoscopic view on selection acting at both the organ and molecular levels. The molecular level is mainly considered in the lecture. The greatest progress to date has been made in relation to the opsin visual pigments. Opsins appeared before eyes did. Two- and three-dimensional organization for rhodopsin in the rod outer segment disk membrane, as well as molecular mechanisms of visual pigments spectral tuning, photoisomerization and also opsin as a G-protein coupled receptor are considered. Molecular mechanisms of visual pigments spectral tuning, namely switching of chromophore (physiological time scale) and amino acid changes in the chromophore site of opsin (evolutionary time scale) is considered in the lecture. Photoisomerization of rhodopsin chromophore, 11-*cis* retinal is the only photochemical reaction in vision. The reaction is extemely fast (less that 200 fs) and high efficient (. is 0.65). The rhodopsin photolysis and kinetics of the earlier products appearance, photo- and bathorhodopsin, is considered. It is known that light is not only a carrier of information, but also a risk factor of damage to the eye. This photobiological paradox of vision is mainly due to the nature of rhodopsin chromophore. Photooxidation is the base of the paradox. All factors present in the phototrceptor cells to initiate free-radical photooxidation: photosensitizers, oxygen and substrates of oxidation: lipids and proteins (opsin). That is why photoprotective system of the eye structures appeared in the course of evolution. Three lines of protective system to prevent light damage to the retina and retina pigment epithelium is known: permanent renewal of rod and cone outer segment, powerful antioxidant system and optical media as cut-off filters where the lens is a key component. The molecular mechanisms of light damage to the eye and photoprotective system of the eye is considered in the lecture. The molecular mechanisms of phototransduction in vertebrates eye is also briefly considered in the lecture. Evolution of vision is an enormous subject for

Mikhail Ostrovsky
Institute of Biochemical Physics, Russian Academy of Sciences, Moscow Joint Institute of Nuclear Research, Dubna and Chair of molecular physiology, Biological Faculty, Lomonosov Moscow State University

Arne T. Skjeltorp, Alexander V. Belushkin (eds.), Evolution from Cellular to Social Scales. 169

thought and investigation. In the postgenomic era evolutionary molecular physiology as a whole and evolutionary molecular physiology of vision can be considered as a key approach for understanding how genome is working.

Keywords Eye, evolution of vision, visual pigments, rhodopsin, opsin, phodopsin photolysis, photoisomerization, phototransduction, visual pigments spectral tuning, photooxidation, light damage, photobiological paradox of vision

1 Evolution of Vision

Let me start my lecture from the Charles Darwin's words about eye that is an "organ of extreme perfection" [Charles Darwin (1809–1882), The Origin of Species by Means of Natural Selection, 1859].

Early organisms evolved photoreceptors more than 600 million years ago. These primitive photoreceptors were capable of signaling light, and that presumably mediated phototaxis. However, it was not until the Cambrian explosion, beginning around 540 million years ago, when image-forming eyes and visual systems appeared. The advantageous capabilities such as sight become crucial to survival. In the various phyla eyes evolved with diverse forms, but it seems that based on certain common underlying features of patterning and development.

The evolution of photoreception, giving rise to eye, offers a kaleidoscopic view on selection acting at both the organ and molecular levels. In studying the evolution of photoreceptors, it would be important to obtain a comprehensive understanding of the evolution of all of the components of the phototransduction signalling cascade. The greatest progress to date has been made in relation to the opsin visual pigments.

Visual information is carried by wavelength, intensity, and/or polarization, which set limits on eye dimensions and detection systems. It is clear from Fig. 1 that visible light is only a narrow part of electromagnetic spectrum. At the earth's surface, about 80% of the sun's energy is restricted to a spectral bandwidth of about 300–1100 nm. This spectral bandwidth has produced the spectral characteristics of a primordial and then in the course of evolution all known visual pigments.

The visual pigment molecules in the photoreceptor cells of the eye constitute the interface between light and organism, between the purely physical and the physiological worlds. Eyes, photoreceptors and visual pigments have been under heavy and changing selection pressures in the course of evolution, as vision has had to adapt to different light environments and to changing behavioral needs of different organisms.

Rod visual pigment rhodopsin was the very first animal membrane protein which primary structure and topography in the membrane, as well which 3D- structure as about 20 years later have been established.

Opsin that is an apoprotein part of rhodopsin molecule is a heptahelical transmembrane protein where seven transmembrane alpha-helical parts ("helical bundles") linked by six extra membrane fragments ("loops") (Fig. 2a).

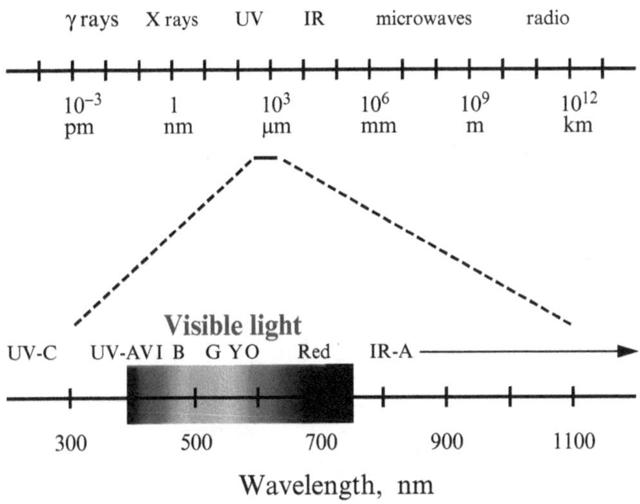

Fig. 1 Electromagnetic spectrum

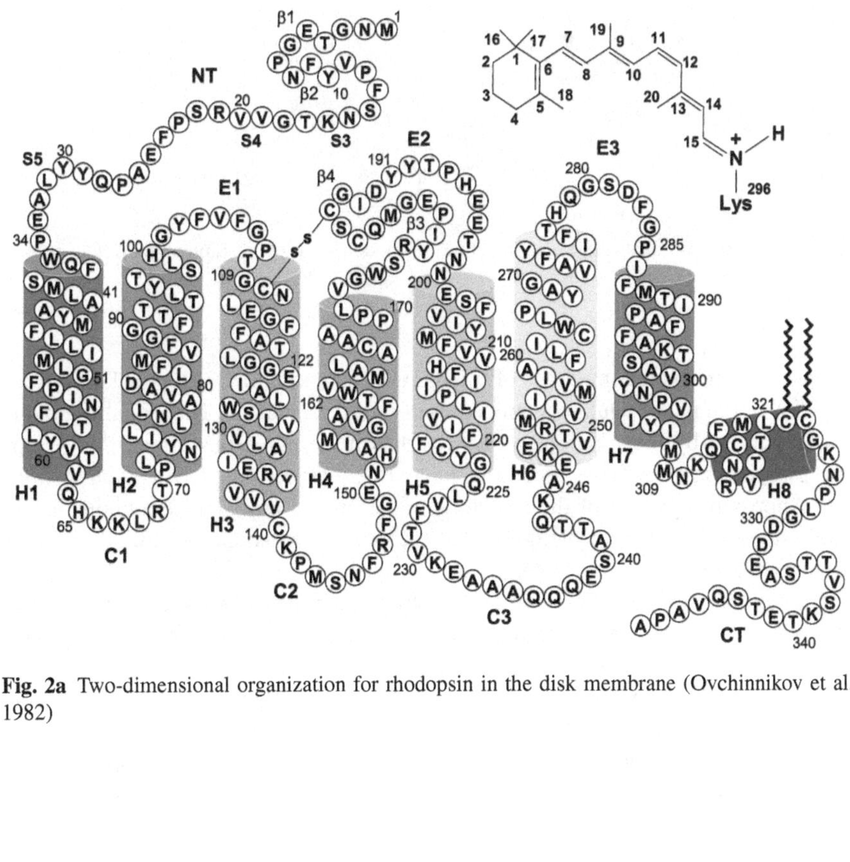

Fig. 2a Two-dimensional organization for rhodopsin in the disk membrane (Ovchinnikov et al., 1982)

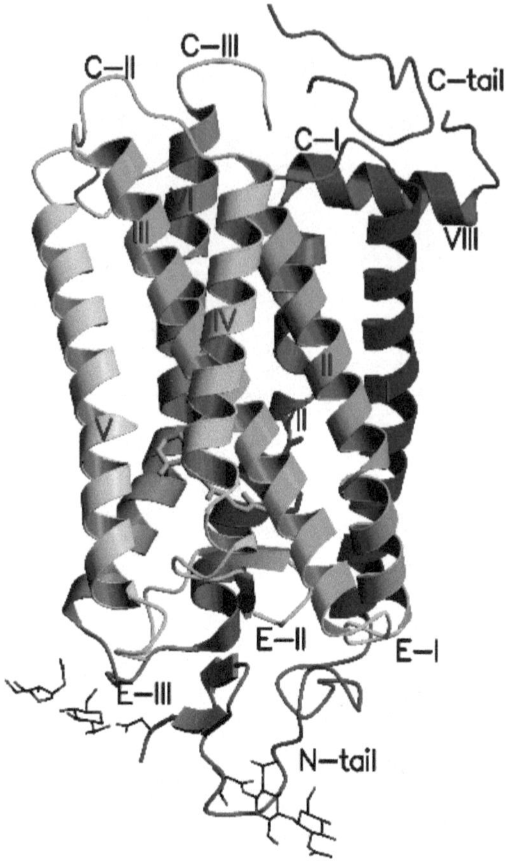

Fig. 2b Three-dimensional (2.8A) organization for rhodopsin in the disk membrane (Palczewski et al., 2000)

Opsin is a protypical member of a large G-protein-coupled receptor family that plays a key role in all regulatory processes of living organisms. The signalling pathways regulated by these receptor proteins determine numerous crucial biological processes, including sensory reception, endocrine regulation and synaptic transmission. Approximately 5% of the human genome (above 600 genes) contains information about of these G-protein-coupled receptor proteins. On the Fig. 2b the 3D structure of rhodopsin is presented.

The picture shows the highly organized structure in the extracellular region forms a basis for the arrangement of the seven-helix transmembrane motif. The chromophore group of all known visual pigments as vertebrates as invertebrates animals as well as human is 11-cis isomer of $retinal_1$ or $retinal_2$ that are aldehydes of vitamin A_1 or A_2. The shift of chromophore in A1 and A2 visual pigments can well be understood as an adaptation to environmental factors, since the A2 chromophore based on 3, 4-dehydroretinal differs from retinal by an extra carbon = carbon

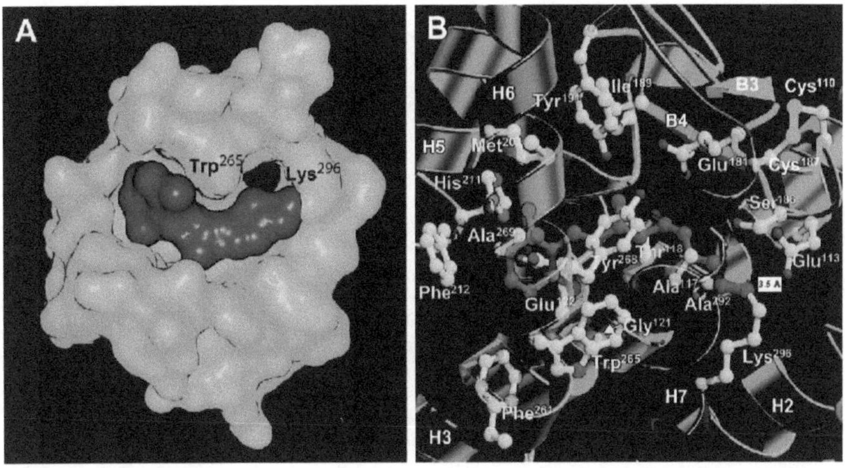

Fig. 3 Rhodopsin: opsin and chromophore group (spectral tunig, photoisomerization, opsin as a G-protein coupled receptor)

double bond, which alters the spectral and thermal properties of the molecule. A shift of chromophore from A1 to A2 broadens the spectrum and red-shifts the pigment. This will also lower the energy barrier for photoactivation (Ea) and make the pigment thermally less stable. Amino acid substitutions in the opsin part of the pigment may alter the charged surrounding around the chromophore, change the energy barrier for photoactivation, and shift the absorbance spectrum accordingly.

The 11-cis isomer is the only isomer that serves as a visual pigments chromophore group among sixteen possible isomers of retinal. Other words, there is very good conformation correspondence between 11-cis retinal and nearby protein (opsin) surrounding in the chromophore domain of rhodopsin.

It is clear to see on the Fig. 3 (left), where 11-cis retinal (red) is situated in the opsin part (green). Next Fig. 3 (right) represents the 3D structure of chromophore site of rhodopsin and chromophore itself that is responsible for, at least, three physiological functions: spectral tuning, 11-cis retinal photoisomerization and keeping rhodopsin as a G-protein-coupled receptor in the dark "silent" state.

Other words, there is very good conformation association between 11-cis retinal and nearby protein (opsin) surrounding in the chromophore domain of rhodopsin.

We have studied recently molecular dynamics of chromophore group, 11-*cis*-retinal within the chromophore site of rhodopsin. Figure 4 shows the molecular dynamics of 11-*cis*-retinal in the chromophore center of the rhodopsin molecule. The picture shows conformational state of the chromophore at the initial ($t = 0$) and the final ($t = 3$ ns) moments. According to our observations, during a short period of time (approximately 300–400 ps) the β-ionone ring rotates about a C6–C7 bond by approximately 60° with respect to the initial configuration of 11-*cis*-retinal.

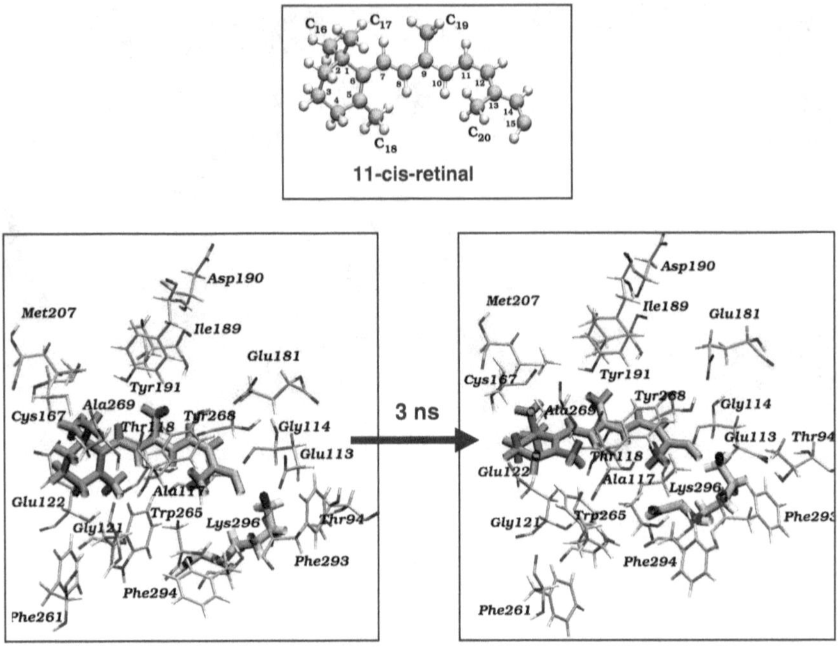

Fig. 4 The molecular dynamics of 11-*cis*-retinal in the chromophore center of the rhodopsin molecule.

Photoisomerization of rhodopsin chromophore, 11-cis retinal is the only photochemical reaction in vision. 11-*cis* retinal is isomerized by light to the all-*trans* isomer, thereby activating the opsin. In case of vertebrate photoreceptors the isomerized retinal is released from opsin and undergoes a complicated cycle (the retinoid cycle) of transport and multiple chemical reactions through which it is isomerized back to the 11-*cis* isomer. Figure 5 (left) shows the rhodopsin photolysis cycle. In fact, this is a basic feature of visual pigment with 11-*cis* retinal as a chromophore group, namely its unusually high rate of chromophore photoisomerization (faster than 200 fs).

The quantum yield of this photoreaction is also high, namely 0.67 or in accordance with recently made estimations –0.65, but not less. The very first photoproduct of rhodopsin photocycle is so called photorhodopsin. Figure 5 (right) shows you the typical experimental result we have obtained in collaboration with the physicists from Institute of Chemical Physics of the Russian Academy of Sciences. Using modern femtoseconds spectroscopy technique we were able to record the appearance of both photorhodopsin (in frame of 200 fs) and then bathorhodopsin. The functional importance of such very high rate of visual pigment chromophore photoisomerization is clear: the energy of absorbed photon should be used as much as possible for useful photochemical reaction that triggers the visual process, and not to be loosed via other concurrent reactions like energy dissipation or fluorescence. The same situation takes place in case of other crucial photobiological

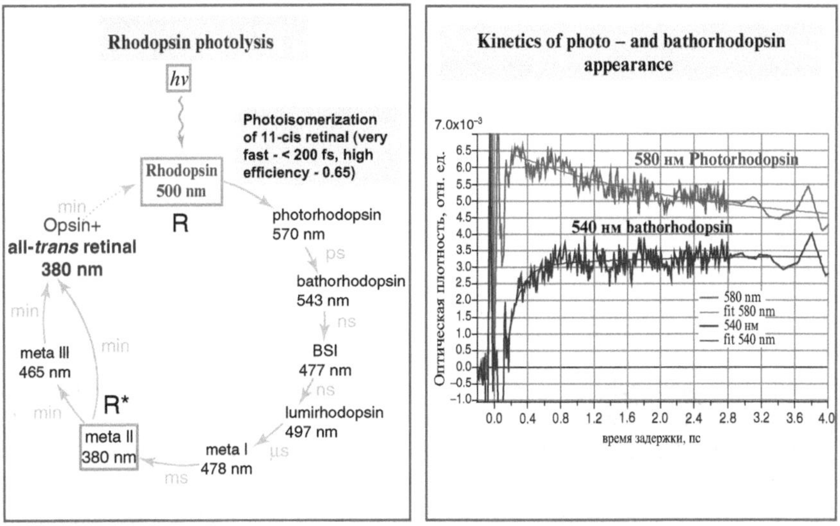

Fig. 5 Photoisomerization of rhodopsin chromophore, 11-cis retinal is the only photochemical reaction in vision

process – photosynthesis, where the biological useful charge separation also occurs in the femtoseconds time scale.

Thus, the structure of chromophore site of visual pigment and its chromophore group is appeared very early and kept conservative in the course of evolution of visual system. As a result the photochemical reaction of vision is able to trigger effectively the process of phototransduction.

From the other hand, variability in the chromophore group protein surrounding and in the structure of chromophore itself (I mean the chromophore can be as aldehyde of vitamin A_1 or vitamin A_2) is responsible for visual pigments spectral tuning. Two time scales can be distinguished in the molecular mechanisms of visual pigments spectral tuning: physiological or ecological time scale and evolutionary time scale. In the first one the spectral tuning is achieved by switching of chromophore from 11-*cis* retinal_1 (aldehyde of vitamin A_1) to 11-*cis* retinal_2 (aldehyde of vitamin A_2) and opposite. In the evolutionary time scale, the spectral tuning is achieved by amino acid changes in the chromophore site of opsin. Speaking about spectral tuning, we should keep in mind that the main functional differences between visual pigments concern two properties: spectral absorbance and thermal stability. The spectral absorbanc refers to the probability of pigment activation as a function of the wavelength of the light (i.e., the energy of the photon). Thermal stability, on the other hand, defines its susceptibility to spontaneous activation by thermal energy alone, which will necessarily degrade the reliability of vision by producing false light-identical signals. To perform well, a visual pigment should catch the photons available in the environment as efficiently as possible while having as low a rate of spontaneous activation as possible. If spectral and thermal properties are

Fig. 6 Opossum shrimp, *Mysis relicta*

coupled, the functional tuning of pigments for high visual sensitivity in different light environments becomes a complex optimization task.

We have studies in collaboration with our finish colleagues from University of Helsinki the question of spectral tuning of visual pigments, and adaptation to different light environments. We have used a group of small crustaceans, *Mysis* as a model. The *Mysis* group of sibling species is very suitable for ecological-evolutionary studies of molecular adaptation. The time of physiological isolation of the different European populations can be accurately dated based on the time of separation of the individual lakes from the sea after the last glaciation, typically 5,000–10,000 years ago. This gives an excellent opportunity to study adaptations and constraints on a "short" and well-defined evolutionary time scale. The "Lake" and "Sea" populations of *M. relicta* have a separation history of only $\sim 9,000$ years. The freshwater lake population of *M. relicta* is adapted to see at extremely dim light, and the sea population is adapted for rather brighter light environment. Figure 6 shows you this small opossum shrimp.

Next Fig. 7 shows you rhodopsin spectra and light transmission in the sea bay and lake.

The absorbance spectra of the visual pigments we have measured by microspectrophotometry technique from two – sea and lake – populations of the *Mysis relicta*. The most remarkable result was the consistent segregation of absorbance maximum values between "Sea" and "Lake" populations: the "Sea" populations were 20–35 nm blue-shifted (see the Fig. 7, left). This result is consistent with the differences in their respective light environment (see the Fig. 7, right). In accordance with our preliminary results this is due to a difference in chromophore. All attempts to find amino acid changes in the opsin part of rhodopsin molecule of "Sea" and "Lake" populations of *M. relicta* were unsuccessful. It means, we have a clear example of physiological time scale spectral tuning. We are studying now if spectral tuning is achieved by switching of chromophore, and not by amino acid substitutions in the opsin. It means that nine-ten thousand years is not enough for evolutionary time scale spectral tuning.

As I said before the spectral tuning of the visual pigments can mainly happen only by two different ways: by changing of the chromphore or by varying the charge environment of the chromophore in the chromphore site of opsin. Let me stress that charge environment of the chromophore is extremely important for

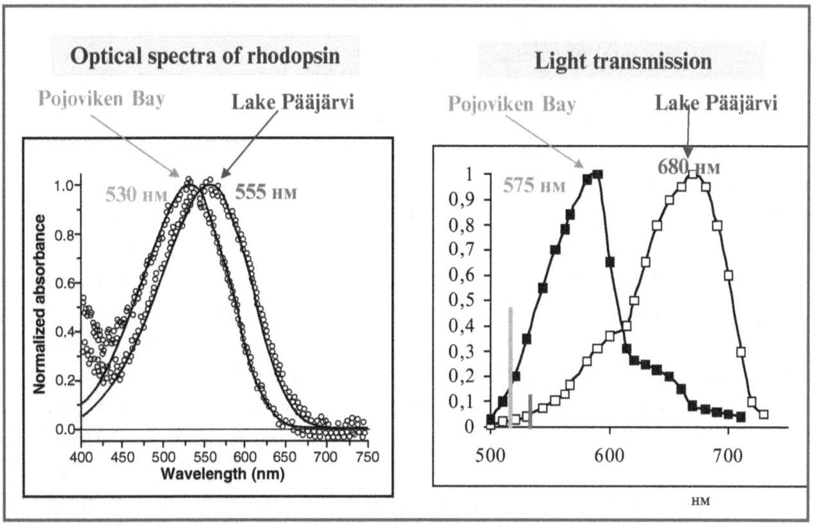

Fig. 7 Rhodopsin spectra and light transmission.

spectral tuning. The chromophore exchange that can take place on a physiological time scale is common for many animals to change the spectral characteristics of their visual pigments during their lifetime. It is well know that the migration from fresh water to seawater is associated with a shift from A2 to A1 – visual pigments. Also, a seasonal change of chromophore is known for many vertebrates like fishes and frogs as well as for invertebrates like crustaceans. Usually, the change from A1 to A2 chromophore will shift to red the absorbance spectrum by approximately 25 nm. In case of two populations of shrimps *Mysis relicta* we have observed a similar.

Now, let me consider the evolutionary time scale spectral tuning achieved by amino acid substitutions in the opsin. It is generally accepted that very early in evolution, namely before the separation of protostomes and deuterostomes, the primordial opsin had already diverged into three main classes: rhabdomeric opsins, so called "photoisomerase" opsins, such as G-protein-coupled receptors, including retinal G-protein-coupled receptor; and ciliary opsins. The later opsins are typical for those photoreceptors in which the visual pigments contain within the expansion of the membrane of a cilium.

Rhabdomeric opsins are typical for most of invertebrate photoreceptor cells and ciliary opsins for vertebrate photoreceptors. The primordial retinal opsin of vertebrates diverged into long-wavelength sensitive (LWS) and short-wavelength-sensitive (SWS) branches, and then the latter split into several sub-groups, each of which is associated with cone-like photoreceptors. I should note that all classes of cone pigment were present before the evolution of the rod pigment, rhodopsin. Other words, it was rather recently discovered that rod-like vertebrate photoreceptors seems to represent the most recent development, most new opsin among ciliary opsins.

Regarding the molecular mechanisms underlying the evolutionary time scale spectral tuning of visual pigments, it should say again that the different absorbance spectra of different visual pigments depend on differences in the electrochemical surrounding of the chromophore, set by the amino acid sidechains of the opsin molecule. The surrounding can be altered by amino acid substitutions from polar to non-polar or from charged to non-charged, changing protein-chromophore interactions. As a result of natural selection of relevant substitutions at critical residues within the opsin chromophore centre, the absorbance band can be fine-tuned over the entire spectrum of visible light, from the one hand, but from the other hand, the correct structural conformation of opsin part for function of the visual pigment like, for example interaction with G-protein (transducin) to initiate the phototransduction cascade in the photoreceptor cell should still be maintained.

Tuning usually involves changes in charge or polarity of the amino acids. The main idea of the evolutionary time scale spectral tuning is: point mutations in the opsin gene, affecting the amino acid sequence of the expressed protein, can affect the spectral characteristics of visual pigments only on an evolutionary time scale. However, in some cases like cichlid fishes spectral "tuning" of the expressed opsin is not achieved through mutations but rather through evolution of gene regulation and differential gene expression. In addition, migrating salmon species lose and regain UV vision through differential gene expression.

In any case, generally speaking, the set of opsin genes on which differential expression may work has originally arisen through amino acid substitutions and natural selection.

The most clearly and well know example of the evolutionary time scale spectral tuning is colour vision.

The common view regarding colour vision based on the numerous data is like this:

- Colour vision evolved very early in vertebrate evolution: about 450–500 million years ago. It was potentially tetrachromatic.
- Four cone opsin classes (long-wave sensitive, LWS, 'green'-sensitive Rh2, short-wave sensitive, SWS2, SWS1) evolved by gene duplication.
- The rod opsin class (Rh1) evolved last through a gene duplication of the 'green'-sensitive (Rh2) cone opsin gene.
- All four cone classes are represented in some modern teleost fish, amphibians, reptiles and birds.
- Mammals have lost two of the cone classes (about 150 million years ago), retaining only the LWS and SWS1 classes. They are generally dichromatic.
- New World monkeys acquired trichromacy through a polymorphism of the LWS gene. Only heterozygous females are trichromatic.
- Old World Primates (and humans) re-evolved trichromacy about 35 million years ago by a gene duplication of the ancestral LWS gene.

The Fig. 8 shows human three cone visual pigments that are responsible for our colour vision.

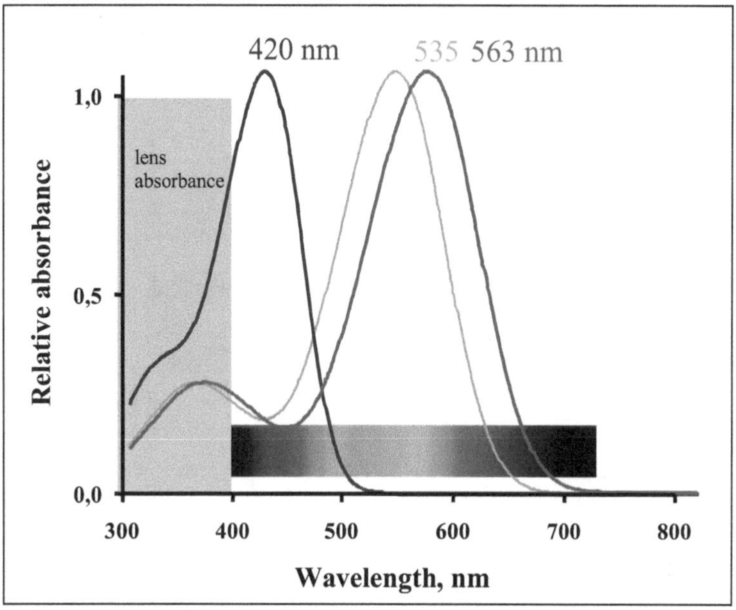

Fig. 8 Human cone visual pigment

In addition, the Fig. 9 illustrates the principle of molecular mechanism of human spectral tuning. It is shown that the spectral difference between the human red and green pigments is determined by amine acid changes at only three tuning sites, namely sites numbers 164, 261 and 269.

Now let me draw your attention to another interesting example of evolutionary time scale spectral tuning. I mean the ionochromic shift that is responsible for

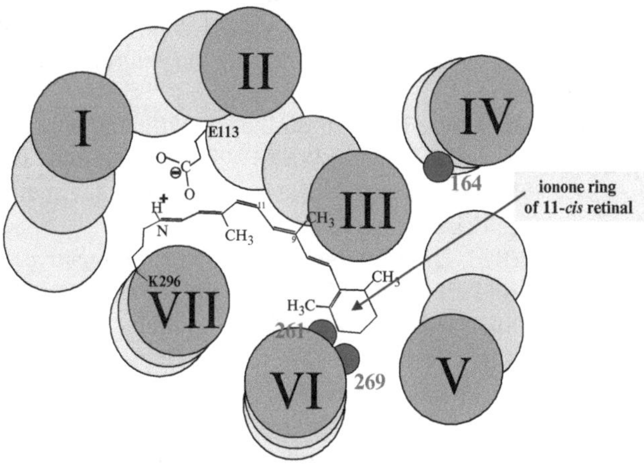

Fig. 9 Molecular mechanisms of visual pigments spectral tuning

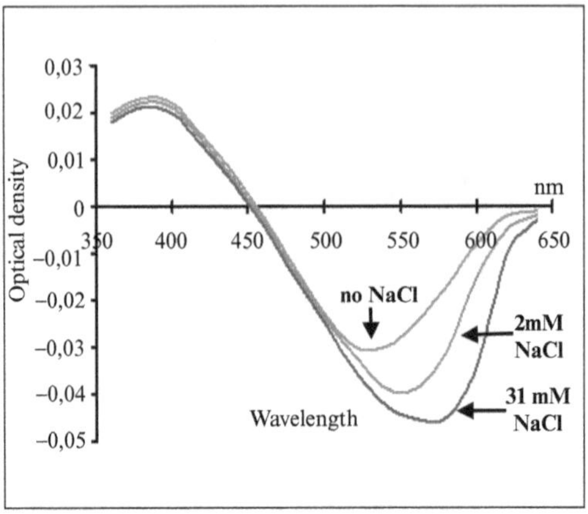

Fig. 10 Blue-shift of chicken long-wave visual pigment (iodopsin) in detergent extract by lowering the chloride ion concentration

long-wave sensitive visual pigment. Amino acid substitutions may affect the charge distribution around the chromophore and thus the position of absorbance maximum not only directly, but also by opening a site for binding an external anion. From this point of view, a notable feature of vertebrate long-wave-sensitive (LWS) visual pigments, those with absorbance maxima longer than about 520 nm, is that their spectral locations are determined not only by the specific amino acid sequences of their opsins, but also by their anionic environment.

It was found at the end of 70th by several authors, as well as by us, that the chicken long-wave visual pigment in detergent extract displaced to shorter wavelengths ('blue'-shift) by lowering the chloride ion concentration. The Fig. 10 shows the phenomenon. A similar 'blue'-shift was also reported for all other vertebrate classes: teleosts, amphibians and mammals.

The Fig. 11 shows you the result of our microspectrophotometric study. As you can see, the long-wave visual pigment of the gecko (*Gecko gecko*) photoreceptors can be displaced to shorter wavelengths by lowering the chloride ion concentration in physiological solution.

Using site-directed mutagenesis to substitute 18 different positively-charged amino acids in the opsins of the human LWS/MWS cone pigments, two residues, histidine at site 197 and lysine at site 200 have been identified as candidates for the chloride-binding site (Fig. 12).

Although the blue-shifted, chloride-free pigment is photosensitive and appears to 'bleach' in a manner similar to normal LWS cone pigments, the question remains as to whether it is functional in terms of its ability to initiate transduction and to generate an electrophysiological response. We have attempted to answer the question by investigating the electrical responses of photoreceptors in isolated goldfish retina perfused with salines containing varying concentrations of chloride. The goldfish (*Carassius auratus*) was selected as a model system since its retina contains four spectrally distinct classes of cone pigment as well as rods.

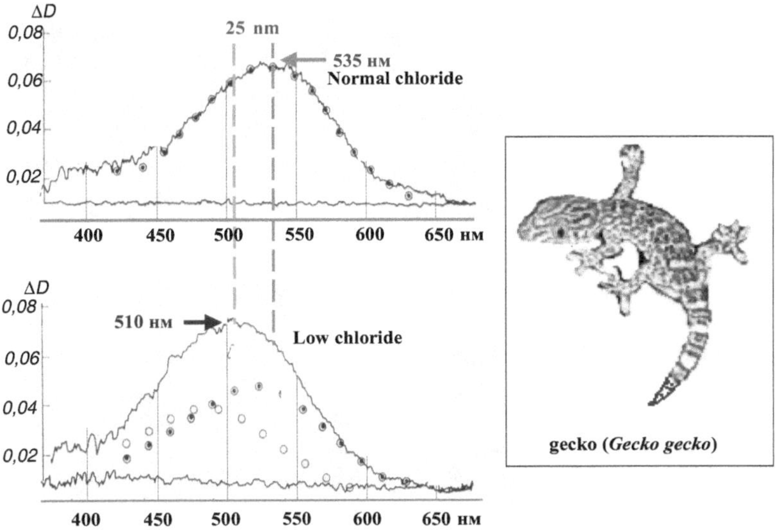

Fig. 11 Blue-shift (25 nm) of red visual pigment in the gecko photoreceptors from 535 to 510 nm by removing chloride from physiological solution

Many fish are tetrachromats. The goldfish as a typical specimen containing four cone visual pigments (Fig. 13). Visual range of goldfish vision from about 300–800 nm, compared to the human range of about 390–740 nm. We have taken advantage of the relatively large spectral difference between the red (LWS) and green (MWS) cone pigments (separated by 80–90 nm). We have compared the behavior of the ionochromic red (LWS) cones with that of the non-ionochromic and green (MWS) cones. The Figure 14 shows you the main results. It is clear to see from comparing the absorbance spectra of goldfish rods and cones obtained by microspectrophotometry, that red (LWS) cones have λmax at 622 nm in normal saline,

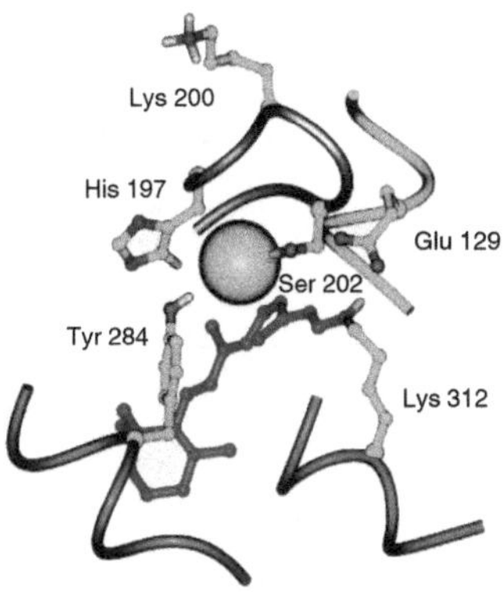

Fig. 12 Model of chloride-binding sites within the chromophore center of LWS pigment of red-sensitive cones

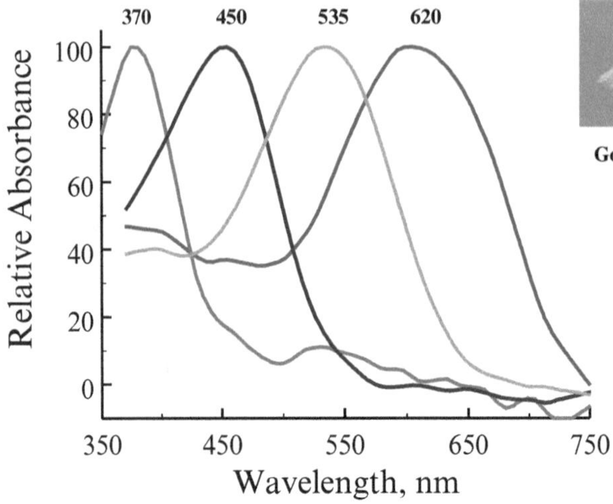

Goldfish, *Carassius auratus*

Fig. 13 Goldfish, *Carassius auratus*: four cone visual pigments

but λmax at 606 nm in chloride-free saline, other words, the shift is quite large (16 nm). At the same time, the absorbance spectra maxima of other cones – UV, blue and green ones, and rods do not shifted (Fig. 14, left). In addition to the spectral shift, the results of the electrophysiological experiments give clear evidence that there is a selective loss in sensitivity of the red (LWS) cones with the removal of chloride ions in comparison with green (MWS) ones (Fig. 14, right).

Based on the results it can be supposed that in the absence of chloride ions the efficiency of transduction in red (LWS) cones is severely reduced, but the green

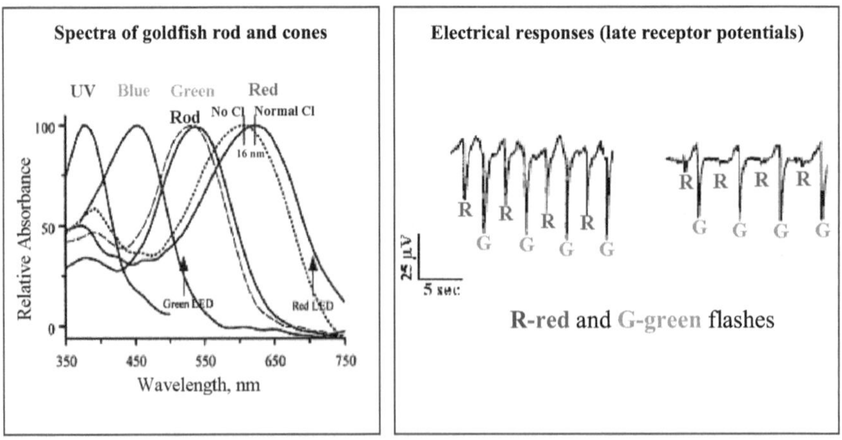

Fig. 14 Goldfish retina: spectral shift and selective inhibition of LWS (red) cones activity after removal of chloride ions from physiological solution

(MWS) and blue (SWS) cones continue to function normally. It should be noted that the shift in λmax and changing in electrical responses in red (LWS) cones are fully reversible upon reintroduction of chloride ions to the medium.

The data obtained by many labs show that the formation of vertebrate LWS visual pigments (λmax $>\sim 520$ nm) cannot be achieved without binding of chloride ions to the chromophore pocket. The importance of the chloride-binding site of LWS visual pigments may not only be to produce a spectral tuning towards longer wavelengths, but it could have a role in stabilizing the molecule in an appropriate conformational state necessary to activate phototransduction.

So, discussing the molecular mechanisms of visual pigments spectral tuning in the evolutionary time scale, one can suggest that a basic feature of the LWS opsins tuning to longer wavelengths different from all other opsin classes is the appearance of a ‘chloride binding pocket’ in LWS opsins’. Although this change has had the benefit of extending the spectral sensitivity of LWS visual pigments further into the long-wavelength region, it has done so at the cost of greatly decreasing the thermal stability of the molecule.

At the end of my lecture, allow me to say that evolution of vision is an enormous subject for thought and investigation. Moreover, we could state now, in the postgenomic era that evolutionary molecular physiology as a whole and evolutionary molecular physiology of vision especially can be considered as a key approach for understanding how genome is working.

Index

Kolibri 360 GmbH
Ferdinand-Jühlke-Straße 7
99095 Erfurt, Deutschland
produktsicherheit@kolibri360.de